Palo Alto Public Library

The individual borrower is responsible for all library material borrowed on his card.

Charges as determined by the CITY OF PALO ALTO will be assessed for each overdue item.

Damaged or non-returned property will be billed to the individual borrower by the CITY OF PALO ALTO.

20M 6/64

❖《❖《❖《 GEORGE ELIOT 》❖》❖》❖

MASTERS OF WORLD LITERATURE

MASTERS OF WORLD LITERATURE SERIES

LOUIS KRONENBERGER, GENERAL EDITOR

George Eliot

❖《《❖《《❖《《 ● 》》❖》》❖》》❖

by Walter Allen

The Macmillan Company, New York
Collier-Macmillan Limited, London

823
E42za

First Printing

The Macmillan Company, New York
Collier-Macmillan Canada, Ltd., Toronto, Ontario

Library of Congress catalog card number: 64-17596

Printed in the United States of America

DESIGNED BY RONALD FARBER

The life of George Eliot has been familiar in broad outline ever since J. W. Cross published his *George Eliot's Life as Related in Her Letters and Journals* in 1885. Professor Gordon S. Haight's great edition of *The Letters of George Eliot* has added vastly to our detailed knowledge of her life and will obviously be the foundation of any future full-length biography. It does not, however, seem to me materially to alter the broad outline of her life, which is all I have space to give here; and where I have used her letters and journals, I have followed Cross. I gladly acknowledge permission to quote granted to me by William Blackwood & Sons, Ltd. I must also thank John Farquharson Ltd. for permission to quote from Henry James's *Daniel Deronda: A Conversation* and Gerald Duckworth and Co., Ltd., for permission to quote from Ann Freemantle's *George Eliot*.

The literature relating to George Eliot and her work is now large. I am conscious of having profited in my interpretation of her life from: *Marian Evans and George Eliot,* by Lawrence and Elisabeth Hanson, *George Eliot and John Chapman,* by Gordon S. Haight, and *Edith Simcox and George Eliot,* by K. A. McKenzie. I have been much helped in my understanding of her art by F. R. Leavis's essays in *The Great Tradition* and his introduction to the Harper Torchbooks edition of *Daniel Deronda, George Eliot: Her Mind and Art,* by Joan

Bennett, *George Eliot: Her Life and Books,* by Gerald Bullett, *George Eliot,* by Robert Speaight, *The Art of George Eliot,* by W. J. Harvey, *Middlemarch from Notebook to Novel,* by Jerome Beaty and the relevant essays in *Early Victorian Novelists,* by David Cecil, *The English Novel: Form and Function,* by Dorothy Van Ghent, *The Victorian Sage,* by John Holloway, and *The Living Novel,* by V. S. Pritchett.

CONTENTS

✦((✦((✦((**GEORGE ELIOT**))✦))✦))✦

·

The Life

1

HOW DID SHE APPEAR to her contemporaries? Four years older, Anthony Trollope was a close friend as well as a fellow-novelist. Unlike her, he was not an intellectual; rather, he was a man of affairs and a man of the world, given to the pleasures of the hunting-field, the card-table and club-life. Both as man and novelist, in the assumptions that governed his conduct and his writing, in his opinions, beliefs, prejudices, the very cast of his mind, he strikes us now as being as representative a mid-nineteenth-century Englishman as could be found. Reviewing his colleagues in the field of fiction in his *Autobiography,* Trollope wrote in 1876:

> At the present moment George Eliot is the first of English novelists, and I am disposed to place her second of those of my time. She is best known to the literary world as a writer of prose fiction, and not improbably whatever of permanent fame she may acquire will come from her novels. But the nature of her intellect is very far removed indeed from that which is common to the tellers of stories. Her imagination is no doubt strong, but it acts in analysing rather than in creating. . . . This searching analysis is carried so far that, in studying her latter writings, one feels oneself to be in company with some philosopher rather than with a novelist. I doubt whether any young person can read with pleasure either *Felix Holt, Middlemarch,* or *Daniel Deronda.* I know that they are very difficult to many that are not young.
>
> Her personsifications of character have been singularly

terse and graphic, and from them has come her great hold
on the public—though by no means the greatest effect which
she had produced. The lessons which she teaches remain,
though it is not for the sake of the lessons that her pages
are read. Seth Bede, Adam Bede, Maggie and Tom Tulliver,
old Silas Marner, and, above all, Tito, are characters which,
once known, can never be forgotten. I cannot say quite so
much for any of those in her later works, because in them
the philosopher so greatly overtops the portrait-painter,
that, in the dissection of the mind, the outward signs seem
to have been forgotten. In her, as yet, there is no symptom
whatever of that weariness of mind which, when felt by
the reader, induces him to declare that the author has
written himself out. It is not from decadence that we do
not have another Mrs. Poyser, but because the author soars
to things which seem to her to be higher than Mrs. Poyser.

Trollope is here writing of George Eliot from the point
of an equal, a fellow-craftsman who knows his own value
and the nature of his achievements. He praises her highly
and handsomely, but not unreservedly; the element of
what he calls the philosopher in her seems to him to
diminish the force of the novelist. She is, it is implicitly
suggested, going beyond the province of the novelist, the
story-teller.

Generous as it is, Trollope's judgment would not have
seemed adequate to many Victorians a generation
younger than himself. As a young don, F. W. H. Myers,
the poet, philosopher and pioneer in psychical research,
invited her in 1873 to visit Cambridge. Years later he
wrote:

I remember how at Cambridge I walked with her once
in the Fellows' Garden of Trinity, on an evening of rainy
May; and she, stirred somewhat beyond her wont, and
taking as her text the three words which had been used so
often as the inspiring trumpet-call of men—the words God,
Immortality, Duty—pronounced with terrible earnestness
how inconceivable was the first, how unbelievable was the

second, and yet how peremptory and absolute the third. Never, perhaps, have sterner accents confirmed the sovereignty of impersonal and unrecompensing Law. I listened, and night fell; her grave, majestic countenance turned towards me like a sibyl's in the gloom; it was as though she withdrew from my grasp, one by one, the two scrolls of promise and left me the third scroll only, awful with inevitable fates. And when we stood at length and parted, amid that columnar circuit of forest trees, beneath the last twilight of starless skies, I seemed to be gazing, like Titus at Jerusalem, on vacant seats and empty halls—on a sanctuary with no Presence to hallow it, and heaven left empty of God.

On another young don of the time, Oscar Browning, she made a comparable impression:

> Although her features were heavy and not well-proportioned, all was forgotten when that majestic head bent slowly down, and the eyes were lit up with a penetrating and lively gaze. She appeared much greater than her books. Her ability seemed to shrink beside her moral grandeur. She was not only the cleverest, but the best woman you had met.

While the historian Lord Acton, a Catholic, wrote of her, an agnostic, on her death: "It seems to me as if the sun had gone out. You cannot think how much I loved her."

The truth is that by the time of her death on December 22, 1880, George Eliot was more than the "mere" storyteller, the "mere" novelist, however distinguished, that Trollope regretted she was not content to be. Like Arnold and Emerson, like Carlyle and Ruskin, her status was that of a special kind of writer, particularly characteristic, perhaps, of the nineteenth century in Britain and the United States, the prophet or sage. In London and Boston alike she was for thousands of readers the maker of the moral law as no novelist before her had been. Hers was a unique position. It was the recognition of this that prompted those among the founders of the London

Library who wished to bar fiction from its shelves to make an exception of hers. Her position at her death can probably be paralleled only by that of Wordsworth a generation or so earlier.

And on the face of it, her ascendancy over her contemporaries is the more remarkable because for many years she was notoriously and publicly living in sin with a married man, ludicrous though the phrase must now sound when applied to her relation with Lewes. The ascendancy was almost as much a triumph of character as of literary and intellectual distinction, of the woman as much as of the novelist. And it was not exercised on intellectuals alone. Complete strangers paid homage to her. Leaving a concert hall with Lewes in 1874, she was approached by an elderly lady who begged to kiss her hand, and a younger followed suit, saying that she was "one of many thousands" who wished to do likewise.

Then, within a matter of thirty years, everything was changed. With the reaction that set in against the nineteenth century in the early days of the twentieth, her reputation slumped; she went the way of the rest of the Eminent Victorians. As recently as 1933 Anne Freemantle could write in her short life of George Eliot:

> . . . even those who write about her to-day are interested in the woman, not the writer, in the author, not in her books. She is no longer even disliked; the gibes about her books as being "dictated to a plain woman by the ghost of Hume," or of her being "George Sand plus science and minus sex," are heard no longer.

George Eliot seemed, in fact, a figure more important in the history of woman's emancipation than in the history of literature.

But that was thirty years ago, and since then the wheel has turned full circle. It is again orthodoxy to say with Trollope, "I am disposed to place her second," even though his first choice among the novelists of his time,

Thackeray, has been ousted by Dickens. During the past two decades, no Victorian novelist except Dickens himself has received more critical attention than George Eliot. Once again, as one surveys the books written on her, one can scarcely escape feeling that she is more than a "mere" novelist, that her novels are in some sense sacred books and that her status remains more like Wordsworth's in the mid-Victorian period than that of any of her contemporaries in the English novel.

The novelist who called herself George Eliot did not do so until she was twenty-eight. Until then, she was Mary Ann—or Marian—Evans. But as Marian Evans she is of great interest in her own right, for she was one of the most remarkable young women of her time. This she would have been if she had never written a line of fiction and become George Eliot. Since she did both and became in consequence one of the two or three most famous Englishwomen of the century, we know a great deal about her, even when she was young and unknown to the world, as may be seen from Dr. Gordon S. Haight's seven-volume edition of her letters, even though this great work may not change nearly as much as is sometimes said the picture of her we get from her husband J. W. Cross's *George Eliot's Life as Related in Her Letters and Journals.*

"Nov. 22, 1819.—Mary Ann Evans was born at Arbury Farm, at five o'clock this morning": so her father, Robert Evans, wrote in his diary. The place and the date are alike significant. Arbury Farm is in the parish of Chilvers Coton, in the north-east of Warwickshire, very close to the border with Leicestershire, in that "central plain" of England "watered at one extremity by the Avon, at the other by the Trent." The words are George Eliot's, in her introduction to *Felix Holt,* in which she describes the countryside as it might have been seen by a traveller in a stage-coach in 1820.

As landscape, this corner of Warwickshire is flat and dull; it is a country in which agriculture and industry have existed cheek by jowl for five hundred years; coal has been mined at Chilvers Coton since the thirteenth century. Indeed, Marian's father, Robert Evans, seems to sum up in himself this double aspect of the countryside. Born in Staffordshire in 1773 and brought up to the family trade of builder and carpenter, he had farmed in Derbyshire before coming to Arbury in 1806 as agent to Francis Newdigate, whose family had been landowners in Warwickshire since the reign of Queen Elizabeth. As Newdigate's agent, he was as much concerned with mining as with agriculture, and even as a little girl Marian must have seen the countryside through his eyes, for it was his habit to take her, "standing between her father's knees as he drove leisurely," on his journeys about the neighbourhood. So that the contrasts of the scene as described in the introduction to *Felix Holt* must have been familiar to her from earliest childhood:

> In these midland districts the traveller passed rapidly from one phase of English life to another: after looking down on a village dirty with coal-dust, noisy with the shaking of looms, he might skirt a parish all of fields, high hedges, and deep-rutted lanes; after the coach had rattled over the pavement of a manufacturing town, the scenes of riots and trades-union meetings, it would take him in another ten minutes into a rural region, where the neighbourhood of the town was only felt in the advantage of a near market for corn, cheese, and hay, and where men with a considerable banking account were accustomed to say that "they never meddled with politics themselves."

The times were troubled ones. The Napoleonic Wars had been ended only four years when Marian was born, and for the middle classes the French Revolution was still a memory vivid in its horror. Politically, it was a time of reaction and repression. Only a few months before Marian's birth, there had taken place in Manchester the "massacre of Peterloo," in which mounted yeomanry, charging a Radical meeting of sixty thousand people, had killed eleven and wounded four hundred. It was Disraeli's England of "the two nations," of which incomparably the larger of the two existed in poverty and without political rights, a country and a period in which those who had feared those who had not, in which for the possessing classes there was a real and ever-present dread of revolution.

Again, it seems fair to assume that the young Marian saw the times through which she passed her childhood through her father's eyes, for Robert Evans dominated her early life and, perhaps more than any other single influence upon her, made her the woman she was. At the beginning and end of her career as a novelist, versions of him appear in her books, as Adam Bede in the novel of that name and as Caleb Garth in *Middlemarch*. He was for her the living embodiment of the principles of duty and responsibility. When, after the publication of *Adam Bede*, someone referred to her as a "self-educated farmer's daughter," she wrote:

> My father did not raise himself from being an artisan to be a farmer: he raised himself from being an artisan to be a man whose extensive knowledge in very varied practical departments made his services valued through several counties. He had large knowledge of building, of mines, of plantations, of various branches of valuation and measurement—all that is essential to the management of large estates. He was held by those competent to judge as *unique* amongst land agents for his manifold knowledge and experience, which enabled him to save the special fees usually paid by landowners for special opinions on the different questions incident to the proprietorship of land.

Great strength of character was married in him with great physical strength. Once, when travelling on the top of a coach, the woman sitting next to him complained that "a great hulking sailor on her other side was making himself offensive." Evans changed places with the woman, seized the sailor by the collar, forced him down under the seat and held him there for the rest of the journey. On another occasion, seeing a couple of labourers waiting for a third to come and help them to move the high heavy ladder used for thatching ricks, he carried the ladder himself without aid from one rick to the other.

A self-made man, he identified himself completely with the landowning class for which he worked. According to George Eliot's husband, J. W. Cross, the French Revolution, which began when Evans was sixteen, "had left an indelible impression on him," and late in life, George Eliot wrote:

. . . Nor can I be sorry, though myself given to meditative if not active innovation, that my father was a Tory who had not exactly a dislike to innovators and dissenters, but a slight opinion of them as persons of ill-founded self-confidence. . . . Naturally enough. That part of my father's prime to which he oftenest referred had fallen on the days when the great wave of political enthusiasm and belief in a speedy regeneration of all things had ebbed, and the supposed [millennial] initiative of France was turning into a Napoleonic empire. . . . To my father's mind the noisy teachers of revolutionary doctrine were, to speak mildly, a variable mixture of the fool and the scoundrel; the welfare of the nation lay in a strong Government which could maintain order; and I was accustomed to hear him utter the word "Government" in a tone that charged it with awe, and made it part of my effective religion, in contrast with the word "rebel," which seemed to carry the stamp of evil in its syllables, and, lit by the fact that Satan was the first rebel, made an argument dispensing with more detailed enquiry.

Marian's life and career were to take paths that Robert Evans could scarcely have conceived, paths that led straight into the most dangerous and nefarious radicalism and dissent. Yet she was her father's daughter to the end, and if a Tory is a man who places the emphasis on the individual's duties whereas the Radical is he who asserts the individual's rights, then she was as much a Tory as her father; and indeed one can see in the passage above a response to Robert Evans's utterances of the words "Government" and "rebel" as much imbued with awe as Myers's to her own pronouncement of the words "God," "Immortality" and "Duty." She herself noted, in the same essay, "Looking Backward," from *Theophrastus Such,* that "certain conservative prepossessions have mingled themselves for me with the influences of our midland scenery, from the tops of the elms down to the buttercups and the little wayside vetches," and these conservative prepossessions were equally mingled with the influence upon her of her father.

At home, however, he does not seem to have been quite the magisterial figure he obviously was in his professional life. According to Cross, he was distinguished by "a general submissiveness in his domestic relations." Cross puts this down to "a certain self-distrust, owing perhaps to his early imperfect education." It is worth noting that self-distrust was a characteristic of his daughter also; but perhaps what is just as important a cause of his "submissiveness" is that, like many another self-made man, he married somewhat above him. Marian's mother, Evans's second wife, came of yeoman stock and was therefore superior socially to her husband's artisan forbears. Cross —and no doubt he had George Eliot's word for this—says there was a "considerable dash of the Mrs. Poyser vein in her" and adds that her family were the prototypes of the Dodsons in *The Mill on the Floss. The Mill on the Floss* is autobiographical in spirit, not in fact, and obviously Robert Evans was no Tulliver. All the same, any man who found himself married to one of the redoubtable Dodson ladies might be excused "a general submissiveness in his domestic relations."

Marian was the third child of this second marriage; Christina had been born in 1814 and Isaac in 1816. Shortly after Marian's birth, Mrs. Evans became something of an invalid, and Christina was sent to school at Attleborough, a mile or two from Griff, the house on the Arbury estate to which the family moved early in 1820. Again without taking the action of *The Mill on the Floss* as a transcription of literal fact, it seems plain that in describing the little girl's hero-worship of her brother, George Eliot was drawing upon her memories of the relations between Marian and Isaac; and in her dependence upon her brother, one sees, as Cross said, "the trait what was most marked in her all through life—namely, the absolute need of some one person who should be all in all to her, and to whom she should be all in all."

During their very early years, apart from attendance at Mrs. Moore's dame's school in a cottage opposite the gates of Griff, the two children seem to have run wild; and the house, grounds and environs of Griff appear to have been a children's paradise. Within the house itself, the great attic was their playroom. Outside, there was the farmyard with its ricks and stables, a copse and, behind the house, a round pool set in and hidden by reeds and willows, a pool from which Marian once surprised herself by hooking a large perch on a line Isaac had lent her. The extreme boundary of the territory they wandered in was the Coventry canal, with its continual procession of coal barges pulled by plodding horses, to stimulate the child's imagination as the symbol of a world of work and activity beyond, of which Griff was the still centre; a symbol duplicated more vividly and romantically twice daily, at ten in the morning and three in the afternoon, by the passing of the Birmingham-Stamford mail-coach, drawn by four greys racing along at ten miles an hour, with scarlet-clad driver above them and scarlet-clad guard, with post-horn, behind. The two images, indeed, of canal and stage-coach, are stamped indelibly on George Eliot's fiction and define its period, which for the most part is that of the generation or so before the coming of the railways changed the face of provincial England and its tempo.

Paradise was lost when she went, at five, to be a boarder at the school at Attleborough where Christina already was; at the same time, Isaac was sent to school at Coventry. Marian came home to Griff occasionally on Saturdays. She told Cross towards the end of her life that her main memory of these early schooldays was of the difficulty she had of getting near the fire in order to keep warm in winter. She was the youngest child in the school by a long way, and the older girls made much of her, calling her "Little Mamma," which suggests anything but high spirits and liveliness in a five-year-old child. She was not, apparently, unhappy, except at nights; then, alone in the dark, she was the prey of night terrors, which continued all her life. She herself believed that this liability to have "all her soul become a quivering fear" was one of the supremely important influences on her future life.

Just as important, however, for her future development was her reading. According to Cross, she had so few books that she read them again and again until she knew them by heart. She rejoiced in *Aesop's Fables* and Joe Miller jest-books. Later, like Maggie Tulliver, she became an addict of Defoe's *History of the Devil*, *The Pilgrim's Progress* and *Rasselas*. But for her the truly seminal author, as one can see now from her fiction and from the temper of her mind, was Scott. She was about seven when a neighbour lent her sister *Waverley*. Marian began to read it, but it was returned before she could finish it. "In her distress at the loss of the fascinating volume," she began to write out the story as far as she had read it herself and continued until her parents were moved to get her the book again. Not only did she discover through him the possibilities of fiction, but his fundamentally Tory imagination also reinforced, as it perhaps humanised, her father's Tory influence on her.

In May, 1826, she went away from home for the first time, on a week's trip with her parents to visit her father's relations in Derbyshire and Staffordshire, a journey that took her into a harder and bleaker country than Warwickshire. It was on this journey, at Wirksworth in Derbyshire, that she met her aunt Elizabeth, her father's brother's wife, a fervent Methodist, who was to be the model of Dinah in *Adam Bede* and the source, too, of the Hetty Sorrel story in that novel.

Marian and her brother were drawing apart. It was, no doubt, the natural order of things, but George Eliot remembered, as "a deeply felt crisis," the time when Isaac was given a pony, to which he became passionately attached, to the exclusion of his sister. She was her father's favourite, as Isaac was her mother's. More even than most children are to their parents, she must have been, in the utterly conventional atmosphere in which she grew up, a bewildering child. Her father, at any rate, recognized her shining intelligence and was proud of it; it was at this time that he seems to have begun taking her with him on his tours of the district. But one sees how her mother, if one equates her with Mrs. Poyser and the Dodson sisters, must have found Marian a trial and a disappointment, for the girl was plain, with a plainness that contained in it no promise of beauty in the future, and with a painful consciousness, moreover, of her plainness, which was to come out later in her fiction in an under-current of disparagement of sexually attractive women. In Mrs. Evans's eyes, she compared poorly with her elder sister, Chrissie, who was pretty, neat, tidy, feminine, a model daughter and the pride of her Dodson-like aunts. At that point of time especially, in that society, Marian's passion for reading must have seemed peculiarly sinister and unnatural.

When she was eight or nine, she was sent with Chrissie to school at the nearby manufacturing town of Nuneaton. She was taught there by Miss Lewis, who became and remained for several years her closest friend, the more important, no doubt, because she was also her first adult friend. She seems to have made no friends among the other girls, and today the intimacy between the child and the mistress would probably be frowned upon as un-healthy. Miss Lewis was an ardent Evangelical church-woman, and her most immediate influence on Marian seems to have been to intensify her religious fervour. One guesses that this was not displeasing to her mother, but it had its dangers, as George Eliot recognised later when she wrote *The Mill on the Floss* and had Philip Wakem upbraid Maggie Tulliver with the words: "You are shut-ting yourself up in a narrow self-delusive fanaticism, which is only a way of escaping pain by starving into dullness all the highest powers of your nature."

When, at the age of thirteen, Marian was removed from Miss Wallinton's, at Nuneaton, to Miss Franklin's school at Coventry—for Robert Evans had realised that the girl deserved the best education it was in his power to give her—her friendship with Miss Lewis continued, and its effects were increased by the tone of the new school. The Misses Franklin were the daughters of a Baptist minister in Coventry, in whom have been found traces of Rufus Lyon, the dissenting minister in *Felix Holt,* and were themselves ardently religious.

According to Cross, Marian "with her chameleon-like nature, soon adopted their religious views with intense eagerness and conviction." She became a leader at prayer-meetings in the school and organized clothing-clubs and charitable activities among the poor of the town. She was, indeed, in the full current of Evangelical religion and was carried away by it as she was always carried away by whatever touched her sense of duty and her imagination. It was, by all accounts, a good school, and Marian was its star pupil. Her English essays "were reserved for the private perusal and enjoyment of the teacher, who rarely found anything to correct." Her music master "reckoned on his hour with her as a refreshment to his wearied nerves, and soon had to confess that he had no more to teach her." When there were visitors to the school, it was Marian who was summoned to play the piano for them; when released, she would rush to her room and "throw herself on the floor in an agony of tears." She must have been forbidding indeed: according to the daughter of a fellow-pupil, the other girls loved her "as much as they could venture to love one whom they felt to be so immeasurably superior to themselves." She was also distinguished from the other girls, and much envied in consequence, by the fact that a weekly cart brought her new-laid eggs from her father's farm.

It was in 1832, the year of the first Reform Act, that Marian Evans went to Miss Franklin's. In the same year in Nuneaton she saw the riot, on the occasion of the North Warwickshire election, which she later recreated in *Felix Holt*. She left Miss Franklin's, at the end of 1835, to go home because of her mother's illness.

Mrs. Evans died in the following summer, Chrissie married soon after, and Marian took over the running of Griff. It was not a period of complete immolation; she was not entirely cut off. A teacher of modern languages came over from Coventry to teach her Italian and German, and so did her music master. She read voraciously; she often went with her father on his rounds, and in the evening she would read or play the piano to him. She ran the home, visited the poor and the sick. But the solitariness and monotony of her life can only have accentuated her tendencies towards introspection and self-scrutiny, her agonised awareness of conscience; and this was not lessened by what was largely her only relief; her letters to Miss Lewis. She told Cross, of this period of her life: "The only thing I should care much to dwell on would be the absolute despair I suffered from of ever being able to achieve anything. No one could ever have felt greater despair . . ." Not unnaturally, she began to suffer from the headaches that from time to time laid her out throughout her life.

Her letters to Miss Lewis at this period are revealing. In August, 1838, she went with Isaac on her first visit to London. One guesses that by now brother and sister were far apart. Isaac had come back from his tutor's at Birmingham a high churchman; it seems that Marian tried to convert him to Evangelicalism, which was suspected of being allied to radicalism and dissent, and Isaac's toryism, like his father's, had only been strengthened by the Nuneaton riots, which threatened everything they identified themselves with. The gulf between them seems to be expressed in the difference in their behaviour on their London trip. "I was not at all delighted with the stir of the great Babel," she wrote to Miss Lewis. Her religious views prevented her from going to theatres with Isaac, and she spent her evenings alone reading. Characteristically, the chief thing she wanted to buy was Josephus's *History of the Jews*. Isaac got it for her and, perhaps no less characteristically, bought himself a pair of hunting sketches.

Later in the year, she wrote to Miss Lewis:

> . . . I have just begun the life of Wilberforce, and I am expecting a rich treat from it. There is a similarity, if I may compare myself with such a man, between his temptations,

or rather *besetments,* and my own, that makes his experience very interesting to me. Oh that I may be made as
useful in my lowly and obscure station as he was in the
exalted one assigned to him! I feel myself to be a mere
cumberer of the ground. May the Lord give me such an
insight into what is truly good, that I may not rest contented with making Christianity a mere addendum to my
pursuits, or with tacking it as a fringe to my garments! May
I seek to be sanctified wholly! My nineteenth birthday will
soon be here (the 22nd)—an awakening signal. My mind
has been much clogged lately by languor of body, to which
I am prone to give way, and for the removal of which I
shall feel thankful.

We have had an oratorio at Coventry lately . . .—the last,
I think, I shall attend. I am not fitted to decide on the
question of the propriety or lawfulness of such exhibitions
to talent and so forth, because I have no soul for music.
. . . I am a tasteless person, but it would not cost me any
regrets if the only music heard in our land were that of
strict worship, nor can I think a pleasure that involves the
devotion of all the time and powers of an immortal being
to the acquirement of an expertness in so useless (at least
in ninety-nine cases out of a hundred) an accomplishment,
can be quite pure or elevating in its tendency.

This is "a narrow self-delusive fanaticism" with a ven-
geance; and, indeed, two years later, at the Birmingham
Musical Festival, she was so wrought upon by the performance that "the attention of people sitting near was
attracted by her hysterical sobbing." But for the time
being, she continued in her letters to Miss Lewis to
escape pain by attempting to starve into dullness all her
highest powers. So we find her, early in 1839, answering
at length an enquiry of Miss Lewis's on the expediency
of reading fiction:

I put out of the question all persons of perceptions so
quick, memories so eclectic and retentive, and minds so
comprehensive, that nothing less than omniverous reading,
as Southey calls it, can satisfy their intellectual man; for

(if I may parody the words of Scripture without profaneness) they will gather to themselves all facts, and heap unto themselves all ideas. For such persons we cannot legislate.

Again, I would put out of the question standard works, whose contents are matters of constant reference, and the names of whose heroes and heroines briefly, and therefore conveniently, describe characters and ideas: such are *Don Quixote*, Butler's *Hudibras, Robinson Crusoe, Gil Blas*, Byron's Poetical Romances, Southey's ditto, etc. Such, too, are Walter Scott's novels and poems. Such allusions as "He is a perfect Dominie Sampson," "He is as industrious in finding out antiquities, and about as successful, as Jonathan Oldbuck," are likely to become so common in books and conversation, that, *always providing* our leisure is not circumscribed by duty into narrow bounds, we should, I think, qualify ourselves to understand. Shakespeare has a higher claim on this for our attention; but we need as nice a power of distillation as the bee, to suck nothing but honey from his pages. However, as in life we must be exposed to malign influences from intercourse with others if we would reap the advantages designed for us by making us social beings, so in books.

Having cleared our way of what would otherwise have encumbered us, I would ask why is one engaged in the instruction of youth to read, as a purely conscientious and self-denying performance of duty, works whose value to others is allowed to be doubtful? I can only imagine two shadows of reasons. Either that she may be able to decide on their desirableness for her pupils, or else that there is a certain power exerted by them on the mind that would render her a more efficient "tutress" by their perusal. . . . I am, I confess, not an impartial member of a jury in this case; for I owe the culprits a grudge for injuries inflicted on myself. When I was quite a little child, I could not be satisfied with the things around me; I was constantly living in a world of my own creation, and was quite contented to have no companions, that I might be left to my own musings, and imagine scenes in which I was chief actress. Conceive what a character novels would give to these

Utopias. I was early supplied with them by those who kindly sought to gratify my appetite for reading, and of course I made use of the materials they supplied for building my castles in the air. But it may be said—"No one ever dreamed of recommending children to read them: all this does not apply to persons come to years of discretion, whose judgments are in some degree matured." I answer that men and women are but children of a larger growth: they are still imitative beings. We cannot (at least those who ever read to any purpose at all)—we cannot, I say, help being modified by the ideas that pass through our minds. We hardly wish to lay claim to such elasticity as retains no impress. We are active beings too. We are each one of the *dramatis personae* in some play on the stage of Life; hence our actions have their share in the effects of our reading. As to the discipline our minds receive from the perusal of fictions, I can conceive none that is benefical but may be attained by that of history.

She concludes:

Religious novels are more hateful to me than merely worldly ones: they are a sort of centaur or mermaid, and, like other monsters that we do not know how to class, should be destroyed for the public good as soon as born. The weapons of the Christian warfare were never sharpened at the forge of romance. Domestic fictions, as they come more within the range of imitation, seem more dangerous. For my part, I am ready to sit down and weep at the impossibility of my understanding or barely knowing a fraction of the sum of objects that present themselves for our contemplation in books and life. Have I, then, any time to spend on things that never existed?

In the same letter, she sends Miss Lewis a copy of some verses she had written on the theme of the scriptural text, "Knowing that shortly I must put off this my tabernacle"; they appeared six months later in the *Christian Observer* over the initials M. A. E., her first published work.

Yet, as other letters to Miss Lewis show, Marian Evans at the age of nineteen was not immersed in "the Christian warfare" to the total exclusion of everything else. Secular things possessed her mind as well. The following indicates the range of her interests at this time:

> . . . my mind, never of the most highly organised genus, is more than usually chaotic; or rather it is like a stratum of conglomerated fragments, that shows here a jaw and rib of some ponderous quadruped, there a delicate alto-relievo of some fern-like plant, tiny shells, and mysterious nondescripts encrusted and united with some unvaried and uninteresting but useful stone. My mind presents just such an assemblage of disjointed specimens of history, ancient and modern; scraps of poetry picked up from Shakespeare, Cowper, Wordsworth, and Milton; newspaper topics; morsels of Addison and Bacon, Latin verbs, geometry, entomology, and chemistry; Reviews and metaphysics,—all arrested and petrified and smothered by the fast-thickening everyday accession of actual events, relative anxieties, and household cares and vexations.

She goes on to mention what must have been a shaping experience:

> I have been so self-indulgent as to possess myself of Wordsworth at full length, and I thoroughly like much of the contents of the first three volumes, which I fancy are only the low vestibule of the three remaining ones. I never before met with so many of my own feelings expressed just as I could like them.

She was in love with learning and perhaps most happy when her learning could be put in the service of Christian warfare, as in the chart of ecclesiastical history she was working on in 1840. The profits from it, when printed, were to go to Attleborough Parish Church:

> The series of perpendicular columns will successively contain the Roman emperors with their dates, the political and religious state of the Jews, the Bishops, remarkable men and events in the several Churches, a column being devoted

to each of the chief ones, the aspect of heathenism and
Judaism towards Christianity, the chronology of the Apos-
tolical and Patristic writings, schisms and heresies, General
Councils, eras of corruption (under which head the remarks would be general), and I thought possibly an
application of the apocalyptic prophecies, which would
merely require a few figures and not take up room. I think
there must be a break in the Chart, after the establishment
of Christianity as the religion of the empire, and I have
come to a determination not to carry it beyond the first
acknowledgement of the supremacy of the Pope by Phocas
in 606, when Mahommedanism became a besom of destruc-
tion in the hand of the Lord, and completely altered the
aspect of ecclesiastical history.

Alas, less than a week later, she is reporting to Miss
Lewis that "Seeley and Burnside have just published a
Chart of Ecclesiastical History, doubtless giving to my
airy vision a local habitation and a name. I console all my
little regrets by thinking that what is thus evidenced to
be a desideratum has been executed much better than if
left to my slow fingers and slower head."

Inevitably today, the Marian Evans of 1840 must
appear a prig; indeed, one critic has said that "nature
and circumstances had conspired" to make her one. In
these letters to Miss Lewis, of course, we see her in an
unusual situation: as the pupil who has become her
teacher's teacher. Up to a point, she may very well be
showing off; and one guesses that Miss Lewis was calcu-
lated to bring out all that was most humourless and
solemn in her. And allowances must be made for the
phraseology of the day and the current cant of Evangelical religion. But much more important than any
priggishness in her at this time is something else. She was
in the grip, this young woman marooned in the wastes of
Warwickshire and conscious of great though as yet un-
specific talents, of what Dr. Johnson called "that hunger
of the imagination which preys incessantly on life."

There was little in her immediate environment, in what for her must then have passed for life, to satisfy her imagination's hunger. At least Evangelicalism was better than nothing. Like all puritan religion, it had its own drama, its own melodrama, and the stage on which this was enacted was the individual soul. It furnished Marian Evans with the essential properties of her mind—and of George Eliot's, too: large conceptions of God, election, duty, responsibility and man's awful destiny that were to remain with her until the end, long after she had ceased to be a Christian, much less an Evangelical Christian.

And this is not all. In the prelude to *Middlemarch* George Eliot wrote, of St. Theresa:

> Theresa's passionate ideal nature demanded an epic life: what were many-volumed romances of chivalry and the social conquests of a brilliant girl to her? Her flame quickly burned up that light fuel, and, fed from within, soared after some illimitable satisfaction, some object which would never justify weariness, which would reconcile self-despair with the rapturous consciousness of life beyond self. . . . That Spanish woman who lived three hundred years ago, was certainly not the last of her kind. Many Theresas have been born who found for themselves no epic life wherein there was a constant unfolding of far-resonant action; perhaps only a life of mistakes, the offspring of a certain spiritual grandeur ill-matched with the meanness of opportunity . . .

That, precisely, was Marian Evans's lot at Griff: her spiritual grandeur was ill-matched by the meanness of her opportunity. The rest of her life as Marian Evans is the story of her successive gropings towards opportunity large enough to match her own generosity and spiritual grandeur, her capacity for greatly feeling and greatly doing.

2

RELEASE FROM THE CONSTRICTIONS of life at Griff came for Marian in March, 1841. Her father, now sixty-seven, decided to retire from business in favour of Isaac, who was just married. Father and daughter moved to Coventry, to a house called Bird Grove, in the Foleshill Road. It meant, almost immediately, an immense widening of Marian's horizon. The Evans's neighbours were Abijah and Elizabeth Pears; Abijah Pears was a man of weight in Coventry and soon to be its mayor. His wife was a former pupil of Mary Franklin. Mary Franklin herself took the news of Marian's arrival in the city to Mrs. Sibree, the wife of a nonconformist minister living in the Foleshill Road, saying that Marian would be "sure to get up to something very soon."

Years later, Mrs. Sibree's daughter recalled their first acquaintance with Marian: ". . . on her first visit to us I well remember she told us of the club for clothing, set going by herself and her neighbour Mrs. Pears, in a district to which she said 'the euphonious name of the Pudding-Pits had been given.'"

That was to be expected; what follows is much more significant:

It was not until the winter of 1841, or early in 1842, that my mother first received (not from Miss Evans's own lips, but through a mutual friend) the information that a total change had taken place in this gifted woman's mind with

respect to the evangelical religion, which she had evidently
believed in up to the time of her coming to Coventry, and
for which, she once told me, she had at one time sacrificed
the cultivation of her intellect, and a proper regard to
personal appearance. "I used," she said, "to go about like
an owl, to the great disgust of my brother; and I would
have denied him what I now see to have been quite lawful
amusements." My mother's grief on hearing of this change
in one whom she had begun to love, was very great; but
she thought argument and expostulation might do much,
and I well remember a long evening devoted to it.

Unavailingly; Marian was probably by now as well
primed in theology and biblical history as the Reverend
John and Mrs. Sibree, and she pretty certainly had a
much better brain than either. It seems that young Miss
Sibree reacted much more enthusiastically to the change
made in Marian by the change in her religious beliefs
than did her parents. This change, she says:

> . . . was traceable even in externals, in the changed tone
> of voice and manner—from formality to a geniality which
> opened my heart to her, and made the next five years the
> most important epoch in my life. She gave me (as yet in
> my teens) weekly lessons in German, speaking freely on
> all subjects, but with no attempt to directly unsettle my
> evangelical beliefs, confining herself in these matters to a
> steady protest against the claim of the Evangelicals to an
> exclusive possession of higher motives to morality—or even
> to religion.

What had happened was simple: Marian had met
Charles Bray and his circle. Bray was thirty, the only son
of a prosperous Coventry ribbon-manufacturer, and, like
his seven sisters, yet another of the Misses Franklin's
former pupils. He had been brought up an Evangelical
but at seventeen was apprenticed in London and saw
much of the family of James Hennell, the brother of

Samuel Hennell, another Coventry ribbon-manufacturer. James Hennell's family was Unitarian and therefore in Evangelical eyes little better than unbelievers. Bray—the behaviour was characteristic—set about converting their minister, and was himself converted.

He returned to Coventry to work for his father an avowed Unitarian. He was the local advanced thinker, and his thinking was even more advanced than, at the time, he was willing to have known: he was reading Shelley's *Queen Mab*, Jeremy Bentham and Jonathan Edwards's *Inquiry into the Freedom of Will*. Bray's public actions as well as his private beliefs led him into conflict with received opinion in Coventry. He founded a school for poor children in a working-class district of the city; since it was nonsectarian, it led to an immediate brush with the Anglican clergy. Later in his career he was to start a newspaper to promulgate his advanced political and social ideas. He was, it is plain, a man whose good humour and great charm went far towards disarming his opponents, and his views did not prevent him from becoming a highly successful business man after his father's death.

In 1836 he married Samuel Hennell's daughter, Charlotte, a Unitarian, as he himself still ostensibly was. But, as he says with the gaiety and good humour that made him so attractive a person:

> The same confidence in what then appeared to me to be the truth, which made me think I could convince the Unitarian Minister, made me now think that I had only to lay my new views on religious matters before my wife for her to accept them at once. Consequently I had provided myself with Mirabeau's *System of Nature*, Volnay's *Ruins of Empire*, and other light reading of that sort to enliven the honeymoon. But again I was mistaken, and I only succeeded in making my wife exceedingly uncomfortable. She had been brought up in the Unitarian Faith, and, as

might be expected in a young person of one-and-twenty, religion with her was not a question of theological controversy or Biblical criticism, but of deep feeling and cherished home associations . . .

Charlotte, indeed, was so disturbed that she applied for help to her brother Charles, who had recently satisfied himself of the truth of Unitarian doctrine. Characteristically, Bray added his persuasions to his wife's, and Hennell agreed to examine the arguments again. The result was his book *Inquiry concerning the Origin of Christianity,* one of the pioneer works in English in what was to be called the Higher Criticism. Since her brother had found that he could no longer accept intellectually even the Unitarian version of Christianity, Charlotte was ready to agree with her husband that they should give up attending church. "This singularity," wrote Bray, "has, I believe, interfered much with my utility in public life."

But Charles Hennell's book had a further consequence: it turned Bray also to authorship; and in 1841, the year in which Marian Evans met him, he published *The Philosophy of Necessity,* the central doctrine of which he stated as follows:

I had one Truth about which I was certain, viz., that no part of the Creation had been left to chance, or what is called free-will; that the laws of mind were equally fixed or determined with those of matter, and that all instinct in beasts, and calculation in man, required that they should be so fixed. I set myself to work, therefore, gradually and laboriously, to build up a system of ethics in harmony with this established fact. I found that *everything* acted necessarily in accordance with its nature, and that there was no freedom of choice beyond this; consequently, if there could be no virtue in the ordinary sense of the term, i.e., in action that is determined, neither could there be any sin . . .

He had been led partly to this conclusion by his discovery a few years earlier of phrenology, a pseudo-science based on the belief that the whole of a man's potential character could be deduced from the shape of his head, and his first book, published in 1839, was on phrenology, to which he remained faithful all his life.

Bray's house, Rosehill, on the outskirts of the city, was much more than merely the centre of local intellectual life. It was, as it were, a station on the underground railway by which advanced ideas spread across the world. You could meet there, as Marian did, Robert Owen, the manufacturer, socialist and philanthropist, of whom she said, "I think if his system prosper it will be in spite of its founder"; the economist and journalist Harriet Martineau; and Ralph Waldo Emerson, who said of Marian, "That young lady has a calm, serious soul." Meeting the Brays, then, brought her into the main current of European ideas and also into the society of men and women dedicated to intellectual honesty and the pursuit of truth.

But it did something else just as valuable to her. One side of life at Rosehill is summed up in Bray's phrase, which was almost a motto: "when the bearskin is under the acacia." Rosehill had a large, tree-shaded lawn, and among these trees was a fine old acacia, the sloping turf about whose roots, according to Bray, made a delightful seat in summer. It was on this turf that the Brays used to spread a large bearskin for their guests. It was a symbol of easy intercourse and uninhibited discussion. Its almost instant effect on Marian was seen by the Sibrees; it was a liberation of spirit, a re-discovery of youth.

Marian met the Brays within two months of coming to live in Coventry, met them at the Pears's next door, for Elizabeth Pears was one of Bray's seven Evangelical sisters. Marian seems to have been impressed by Bray from the beginning, but, disconcertingly, she was not to meet him again for several months. She had, in fact, reminded him by her tone of conversation and manner too much of the seven Evangelical sisters, while Elizabeth seems to have conceived it her duty not to encourage friendship between the brilliant young woman next door, however ardent her Evangelicalism, and her infidel brother, whose powers of persuasion were well known. Elizabeth changed her mind, however; it struck her that Marian might be the very person to win Bray back to Christianity, and on November 2, 1841, Marian set out for Rosehill with precisely this purpose in mind, though not with any great confidence. Before going, she wrote to Miss Lewis: "I am going, I hope, today to effect a breach in the thick wall of indifference behind which the denizens of Coventry seem inclined to intrench themselves; but I fear I shall fail."

She was, in fact, already much less firmly convinced an Evangelical than anyone knew, for, when she met Bray, it came out in conversation that she had read Hennell's *Inquiry,* which is something that she had certainly kept hidden both from Mrs. Pears and from Miss Lewis. Bray had forgotten that he had met her before and was struck by her intelligence, her humour, her immediate response to his own wit; above all, by her phrenological development. "We became," he wrote, "friends at once."

She does not seem to have made any determined effort to convert him. He asked her whether she had read Carlyle's *Sartor Resartus,* which she had, and Emerson's *Essays,* which she had not. She admitted to having read the *Inquiry;* and she left Rosehill with *The Philosophy of Necessity* under her arm. Within less than a fortnight, she was writing to Miss Lewis:

> My whole soul has been engrossed in the most interesting of all inquiries for the last few days, and to what result my thoughts may lead, I know not—possibly to one that will startle you; but my only desire is to know the truth, my

only fear to cling to error. I venture to say our love will not decompose under the influence of separation, unless you excommunicate me for differing from you in opinion. Think—is there any *conceivable* alteration in me that would prevent your coming to me at Christmas?

Whether Miss Lewis went to Foleshill for Christmas and how she reacted to Marian's defection not merely from Evangelical Christianity but from Christianity itself, we do not know. For Marian herself, the discarding of her old beliefs seems to have been like stepping out of the husks of a chrysalis. She was, as she said years later, "in a crude state of free-thinking"; but, in fact, change in belief had not altered her moral certainties. There was, of course, the problem of her adjustment to her new principles, and it is plain that for several months she found its solution agonisingly difficult.

She decided that she could no longer attend church, but she was caught in an intolerable dilemma. There was, on the one hand, her duty to her conscience; on the other, the depth of her love for her father. In Robert Evans's eyes, her behaviour was unforgivable; she was outraging everything that to him was sacrosanct by tradition and right conduct, and he made it a condition of her continuing to live with him that she go to church. When she still refused, he put the lease of his house into an agent's hands, planning to go to live with Chrissie. Marian thought of going into lodgings in Leamington and finding a job as a teacher, but nothing came of it, and for nearly two months she continued to live with her father in Foleshill Road.

During this period she wrote to Miss Lewis: "I have had a weary week. At the beginning more than the usual amount of *cooled* glances, and exhortations to the suppression of self-conceit. The former are so many hailstones that make me wrap more closely around me the mantle of determinate purpose: the latter are needful, and have a tendency to exercise forbearance, that well repays the temporary smart." Then, at the end of February, she was packed off to her brother Isaac and his family at Griff, to stay there until she had made up her mind.

She did; and one assumes that she did so partly under the influence of the scenes of her childhood. In March she wrote, to Mrs. Pears, a letter that strikes a new note in her development:

> I have here in every way abundant and unlooked-for blessings—delicacy and consideration from all whom I have seen; and I really begin to recant my old belief about the indifference of all the world towards me, for my acquaintances of this neighbourhood seem to seek an opportunity of smiling on me in spite of my heresy. All these things, however, are but the fringe and ribbons of happiness. . . .
>
> I am more and more impressed with the duty of *finding* happiness. On a retrospection of the past month, I regret nothing so much as my own impetuosity both of feeling and judging. I am not inclined to be sanguine as to my dear father's future determination, and I sometimes have an intensely vivid consciousness, which I only allow to be a fleeting one, of all that is painful and that has to be. I can only learn that my father has commenced his alterations at Packington, but he only appears to be temporarily acquiescing in my brother's advice "not to be in a hurry." I do not intend to remain here longer than three weeks, or, at the very farthest, a month; and if I am not then recalled, I shall write for definite instructions. I must have a *home,* not a visiting place. I wish you would learn something from my father, and send me word how he seems disposed.

She returned to Foleshill Road; her father, persuaded by Isaac, Miss Franklin and the Brays also, received her back—and she resumed church-going. Why, she explained eighteen months later in a letter to Sara Hennell, Charlotte Bray's sister, who was to be, with Bray and his

wife, one of her closest friends for many years. It is an
important letter, for in it Marian enunciates the doctrine
of what she calls "truth of feeling," which she was never
to depart from:

The subject of your conversation with Miss D. is a very
important one, and worth an essay. I will not now inflict
one of mine on you, but I will tell you, as briefly as possible,
my present opinion, which you know is contrary to the one
I held in the first instance. . . . The first impulse of a young
and ingenuous mind is to withhold the slightest sanction
from all that contains even a mixture of supposed error.
When the soul is just liberated from the wretched giant's
bed of dogmas on which it has been racked and stretched
ever since it began to think, there is a feeling of exultation
and strong hope. We think we shall run well when we have
the full use of our limbs and the bracing air of independence, and we believe that we shall soon obtain something
positive which will not only more than compensate for
what we have renounced, but will be so well worth offer-
ing to others, that we may venture to proselytise as fast as
our zeal for truth may prompt us. But a year or two of
reflection, and the experience of our own miserable weakness, which will ill afford to part even with the crutch of
superstition, must, I think, effect a change. Speculative truth
begins to appear but a shadow of individual minds. Agree-
ment between intellects seems unattainable, and we turn
to the *truth of feeling* as the only universal bond of union.
We find that the intellectual errors which we once fancied
were a mere incrustation have grown into the living body,
and that we cannot in a majority of cases wrench them
away without destroying vitality. We begin to find that
with individuals, as with nations, the only safe revolution
is one arising out of the wants which their own progress
has generated. It is the quackery of infidelity to suppose
that it is a nostrum for all mankind, and to say to all and
singular, "Swallow my opinions, and you shall be whole."
If, then, we are debarred by such considerations from trying to reorganise opinions, are we to remain aloof from our

fellow-creatures on occasions when we may fully sympathise with the feelings exercised, although our own have been melted into another mould? Ought we not on every opportunity to seek to have our feelings in harmony, though not in union, with those who are often richer in the fruits of faith, though not in reason, than ourselves? The results of nonconformity in a family are just an epitome of what happens on a larger scale in the world. An influential member chooses to omit an observance which, in the minds of all the rest, is associated with what is highest and most venerable. He cannot make his reasons intelligible, and so his conduct is regarded as a relaxation of the hold that moral ties had on him previously. The rest are infected with the disease they imagine in him. All the screws by which order was maintained are loosened, and in more than one case a person's happiness may be ruined by the confusion of ideas which took the form of principles. But, it may be said, how then are we to do anything towards the advancement of mankind? Are we to go on cherishing superstitions out of a fear that seems inconsistent with any faith in a Supreme Being? I think the best and the only way of fulfilling our mission is to sow good seed in good (i.e., prepared) ground, and not to root up tares where we must inevitably gather all the wheat with them. We cannot fight and struggle enough for freedom of inquiry, and we need not be idle in imparting all that is pure and lovely to children whose minds are unbespoken. Those who can write, let them do so as boldly as they like—and let no one hesitate at proper seasons to make a full *con*fession (far better than *pro*fession). St. Paul's reasoning about the conduct of the strong to the weak, in the 14th and 15th chapters of Romans, is just in point. . . .

This is surely a remarkable letter for anyone of twenty-four to have written; indeed, one is tempted to say that only George Eliot could have written it, with its deep sense of natural piety, its balance of the claims of intellectual radicalism and traditional values.

Marian had returned to church-going, but in assuming

that this meant a return to belief, her pious neighbours the Pearses and the Sibrees and her old teachers the Misses Franklin had let themselves in for any number of shocks. The Misses Franklin sent a well-known local Baptist minister to Foleshill to re-convert her; he retired hurt, exclaiming plaintively: "That young lady must have had the devil at her elbow to suggest doubts, for there was not a book that I recommended to her in support of Christian evidences that she had not read." The Reverend Mr. John Sibree himself then had a go, and his daughter remembered, years later (to go back to her reminiscences of Marian, quoted above), "her indignant refusal to blame the Jews for not seeing in a merely spiritual Deliverer a fulfillment of promises of a temporal one; and her still more emphatic protest against my father's assertion that we had no claim on God."

But the Sibrees were not the people to admit failure; they had heavier guns in reserve and fetched up to confute her a Mr. Watts, a theologian on the staff of the Baptist college at Birmingham. He was learned in the Higher Criticism, had read Strauss's *Leben Jesu,* could meet her point to point; but he retired, saying: "*S*he has gone into the question." Undaunted, Mrs. Sibree kept up the pressure—to no avail; and soon Miss Sibree was writing to her brother, a theological student at Halle University, that Marian "seems more settled in her views than ever, and rests her objections to Christianity on this ground, that Calvinism is Christianity, and this granted, that it is a religion based on pure selfishness."

One would be less than fair to good people if one did not stress the genuine friendship in which the Pearses, the Sibrees and the rest held Marian. She was a phenomenon they had not met before: her goodness was self-evident, and yet they could only believe that hers was a soul perilously in the balance. There must have been tact and forebearance exercised on both sides. When

Mary Sibree herself began to doubt her Evangelical faith in 1843, it was to Marian she turned for sympathy and advice. The advice, to judge from one letter, was excellent: "Mrs. D's mother is, I dare-say, a valuable person; but do not, I beseech thee, go to old people as oracles on matters which date any later than their thirty-fifth year. Only trust them, if they are good, in those practical rules which are the common property of long experience."

The Sibrees and the rest were powerless against Marian's honesty of intellect now that it had found assurance in the example of the Brays and the Rosehill circle. As Mary Sibree wrote: "Mr. and Mrs. Bray and Miss Hennell, with their friends, were *her* world—and on my saying to her once, as we closed the garden door together, that we seemed to be entering a Paradise, she said, 'I do indeed feel that I shut the world out when I shut that door.'"

Today, such language seems too gushing, too ecstatic. Yet the contrast between what she found at Rosehill and what she had known in her life before can hardly be exaggerated. It must have been as though she had stumbled into a Thélème of a kind, however un-Rabelaisian. She found there delight in music and conversation and the pursuit of truth. It was at Rosehill that she discovered Spinoza, whose *Tractatus theologico-politicus* she began to translate. Spinoza did much to reconcile her to the bleakness of Bray's beliefs, for he added the dimension of poetry that was always necessary for her. He was, she thought, Wordsworthian. And above all, at Rosehill she found friends with whom she could be an equal among equals: Bray, his wife and his sister Sara, to all of whom she responded with different sides of her complex nature. When the Brays went on holiday, she went with them, touring the English Midlands with them, accompanying them to Wales, the Lake District, London

and Scotland. From Bray's autobiography we can see
how she struck them:

> I saw a great deal of her, we had long frequent walks
> together, and I consider her the most delightful companion
> I have ever known; she knew everything. She had little
> self-assertion; her aim was always to show her friends off
> to advantage—not herself. She would polish up their wit-
> ticisms, and give them the *full* credit for them. But there
> were two sides; hers was the temperament of genius which
> has always its sunny and shady sides. She was frequently
> very depressed—and often very provoking, as much as she
> could be agreeable—and we had violent quarrels; but the
> next day, or whenever we met, they were quite forgotten,
> and no allusion made to them. Of course we went over all
> subjects in heaven or earth. We agreed in opinion pretty
> well at that time, and I may claim to have laid down the
> base of that philosophy which she afterwards retained.

His description of her temperament, based on the
findings of phrenology—a cast of her head was made in
1844—as much as on his own observation, is equally
interesting and largely substantiated by the course of her
life:

> She was of a most affectionate disposition, always re-
> quiring someone to lean upon, preferring what has hitherto
> been considered the stronger sex, to the other and more
> impressible. She was not fitted to stand alone. Her sense of
> Character—of men and things, is a predominantly intellec-
> tual one, with which the Feelings have little to do, and
> the exceeding fairness, for which she is noted, towards all
> parties, towards all sects and denominations, is probably
> owing to her little feeling on the subject,—at least not
> enough to interfere with her judgment. She saw all sides,
> and they are always many, clearly and without prejudices.

It was in 1843 that Marian met Dr. Brabant. She was
one of the bridesmaids at the wedding in London be-

tween his daughter Rufa and Charles Hennell. After the
wedding "she afterwards paid a visit to Dr. Brabant,
at Devizes," to quote Bray, "in order to cheer him upon
the loss of his only daughter." Brabant was a man of
sixty-three, with a wife and also a sister-in-law living
with him; Marian was twenty-four. Brabant was a wealthy
man, a physician who had given up his practice to devote
himself to biblical scholarship.

According to the novelist and journalist Eliza Lynn
Linton, who was later, for a time, much caught up in
Marian's life, though hardly as a dispassionate observer,
Brabant "was a learned man who used up his literary
energies in thought and desire to do rather than in actual
doing, and whose fastidiousness made his work something like Penelope's web. Ever writing and re-writing,
correcting and destroying, he never got further than the
introductory chapter of a book which he intended to be
epoch making, and the final destroyer of superstition and
theological dogma." In this respect, certainly, we get a
glimpse of a man not unlike Mr. Casaubon in *Middlemarch*. He was unlike Casaubon, however, in being a
German scholar; and when he read Charles Hennell's
Inquiry, he sent it to his friend Strauss, the author of
Leben Jesu, which had appeared three years earlier.
Strauss was as much impressed by Hennell's book as Brabant had been, arranged for its translation into German
and wrote a eulogistic preface to it.

Brabant had similarly arranged for an English version
of *Leben Jesu*, and his daughter Rufa was working on
the translation when she met and married Hennell.
Someone had to take her place. It was to be Marian.
Without question, the work was one of great importance, intellectually worthy of her.

What exactly happened when Marian got to Devizes is
not certain. Cross, George Eliot's official biographer,
glides over the episode in three lines; but it seems clear

that Marian was quite bowled over by Brabant. She began to study Greek with him and read to him in German. He called her Deutera, because she was to be a second daughter to him, and within a week of her arrival she was writing to Sara: "I am in a little heaven here, Dr. Brabant being its archangel." It seems that Caroline Bray tried to warn her. She replied that she was "never weary of his company." According to Eliza Lynn Linton:

> A family tradition chronicles a scene which took place between the young woman and the elderly man, when she knelt at his feet and offered to devote her life to his service. Between a wife who, though blind, counted for something in the councils of the household, and a vigilant sister-in-law who looked sharply after the interests of all concerned, this offer of a life-long devotion proved abortive. The enthusiasm of the girl was promptly stifled under the wet blanket thrown over it by an alarmed wife and a sister who thought such spiritual attachments might lead to danger; and Mary Ann Evans left the house a sadder woman than she entered it.

When allowance is made for the plain bitchiness that informs the passage, it still seems clear that Marian was indiscreet. Brabant's daughter, Rufa Hennell, herself told John Chapman, according to his diary of 1851, that Marian

> . . . in the simplicity of her heart and her ignorance of (or incapability of practising) the required conventionalisms, gave the Doctor the utmost attention; his Sister-in-law, Miss S. Hughes, became alarmed, made a great stir, excited the jealousy of Mrs. Brabant. Miss Evans left. Mrs. B. vowed she should never enter the house again, or that if she did, she, Mrs. Brabant, would instantly leave it. Mrs. Hennell says Dr. B. ungenerous and worse towards Miss E., for though he was the chief cause of all that passed, he acted towards her as though the fault lay with her alone.

His unmanliness in the affair was condemned more by Mrs. Hennell than by Miss E. herself when she (a year ago) related the circumstances to me.

For us now, the episode has a twofold interest. It brings out, as nothing we know of her life earlier does, what may be called the Maggie Tulliver side of her character. However well-disciplined and "masculine" her mind might be, she was at the mercy of her emotions, perhaps at the mercy of her need for affection and self-sacrifice. In some ways, as her biography shows, she was a born victim, for she was to repeat a similar pattern of behaviour with Chapman a few years later. But the episode also provides us—in part, at any rate—with the genesis of Casaubon in *Middlemarch*. When asked by a friend to tell on whom the character was based, "with a humorous solemnity, which was quite in earnest, she pointed to her own heart." As Dr. Gordon S. Haight conjectures: "There in her pain and humiliation lay the venom that was to give Casaubon his horrible vividness years later."

She went back to Coventry to translate *Leben Jesu*. It took her three years, during which time, of course, she continued to keep house for her father, besides visiting the poor, teaching in an industrial school, keeping up her lessons in classical and modern languages and going to lectures at the Mechanics' Institute. In addition to all this, she taught herself Hebrew, in order to know Strauss's sources at first hand. To begin with, she averaged six pages of translation a day, a rate which, if kept up, would have seen the work finished within a year. It was not kept up; headaches interrupted her, and so did the excursions organised by the irrepressible Bray and the distractions of Rosehill. Her father, too, was often ill and relied more and more on her; she read to him the novels of Walter Scott.

Nor was the task of translating Strauss altogether congenial. She might admire his thoroughness but she often found he outraged the "truth of feeling." "I am never pained," she wrote to Sara Hennell, "when I think Strauss right—but in many cases I think him wrong, as every man must be in working out in detail an idea which has general truth, but is only one element in a perfect theory—not a perfect theory in itself." She worked with an ivory image of Christ on the Cross above her desk. In February, 1846, we find Mrs. Bray writing to Sara that Miss Evans

> . . . says she is Strauss-sick—it makes her ill dissecting the beautiful story of the crucifixion, and only the sight of the Christ image and picture make her endure it. Moreover, as the work advances nearer its public appearance, she grows dreadfully nervous. Poor thing, I do pity her sometimes, with her pale sickly face and dreadful headaches, and anxiety too about her father. This illness of his has tried her so much, for all the time she had for rest and fresh air she had to read to him. Nevertheless she looks very happy and satisfied sometimes in her work.

Then, in the middle of April, the translation was finished. "As her first enjoyment," Mrs. Bray wrote to Sara, Marian "means to come and read Shakespeare through to us." A month or so later, she went to join Sara for a holiday in London, and a letter to Mrs. Bray, who was to join them, gives some idea of the relief and the exhilaration she felt now that she was free of the burden of Strauss:

> I cannot deny that I am very happy without you, but perhaps I shall be happier with you, so do not fail to try the experiment. We have been to town only once, and are saving all our strength to "rake" with you; but we are as ignorant as Primitive Methodists about any of the amusements that are going. Please to come in a very mischievous,

unconscientious, theatre-loving humour. . . . Don't bring us any bad news or any pains, but only nods and becks and wreathed smiles.

In London, she met her publisher, John Chapman, for the first time; and her translation (anonymous) of *Leben Jesu* appeared on June 15, 1846.

3

ROBERT EVANS DIED on the night of May 31, 1849. A few hours earlier, Marian had written to the Brays: "What shall I do without my father? It will seem as if part of my moral nature were gone." They had lived together alone for eight years, and during the last months of his illness she had nursed him alone, finding relief in translating Spinoza's *Tractatus theologico-politicus.*

Intellectually, they can have had nothing to say to each other; she lived in a world of ideas quite outside his comprehension. Yet this is to see the relationship between them in the most superficial way. What he meant to her is indicated in the sentence following that in the letter to the Brays quoted above, a sentence Cross omitted in his transcription of the letter in his life of George Eliot: "I had a heroic vision of myself last night becoming earthly, sensual, and devilish for want of that purifying, restraining influence."

As ever, in this immediate situation of bereavement and despair, the Brays came to the rescue. They were going abroad and invited Marian to join them. She did; and on June 11 they left, going by way of Paris, Lyons, Avignon, Marseilles, Nice, Genoa, Milan, Como, Lake Maggiore and Chamonix to Geneva, where they arrived in the third week of July. There, Marian decided to stay on alone, saying, "I will never go near a friend again until I can bring joy and peace in my heart and in my face."

She settled in at a *pension*. French, Germans, Americans were among the guests as well as English people, and her letters show how livelily she responded to this new environment:

"The American lady embroiders slippers,—the mama looks on and does nothing. The Marquis and his friends play at whist; the old ladies sew; and Madame says things so true that they are insufferable."

"The Marquis is the most well-bred, harmless of men. He talks very little—every sentence seems a terrible gestation, and comes forth *fortissimo;* but he generally bestows one on me, and seems especially to enjoy my poor tunes. . . . The grey-headed gentleman got quite fond of talking philosophy with me before he went. . . . The young German is the Baron de H. I should think he is not more than two or three and twenty, very good-natured, but a most determined enemy to all gallantry. I fancy he is a Communist; but he seems to have been joked about his opinions by Madame and the rest, until he has determined to keep a proud silence on such matters."

"You would not know me if you saw me. The Marquise took on her the office of *femme de chambre* and drest my hair one day. She has abolished all my curls, and made two things stick out on each side of my head like those on the head of the Sphinx. All the world says I look infinitely better; so I comply, though to myself I seem uglier than ever—if possible."

Alas, contentment did not last long. She was prostrated by headaches, "prolonged, in fact, by the assiduities of the good people"; and it is clear from the letters that feelings of guilt were beginning to work in her: "This place looks more lovely to me every day . . . one can hardly believe one's self on earth: one might live here and forget that there is such a thing as want or labour or sorrow. The perpetual presence of all this beauty has somewhat the effect of mesmerism or chloroform. I feel sometimes as if I were sinking into an agreeable state of numbness on the verge of unconsciousness."

Then, guests at the *pension* came and went just as she was beginning to know them. She was worried, too, about money, for the income her father had left her was small. "Do you think," she writes to the Brays, "any one would buy my 'Encyclopaedia Britannica' at half-price, and my globes? If so, I should not be afraid of exceeding my means, and I should have a little money to pay for my piano, and for some lessons of different kinds that I want to take."

Still, she would not leave Geneva, and in October she took an apartment in the house of M. and Mme. d'Albert, "middle-aged-musical, and I am told, have *beaucoup d'esprit.*" M. d'Albert was the curator of the Athenée, the art gallery in Geneva; later, he was to translate *Adam Bede, Silas Marner, Romola* and *Felix Holt* into French. The relationship Marian slipped into with the d'Alberts was of a kind that she was, all the evidence shows, always unconsciously seeking, the kind she had with the Brays and had failed to achieve with the Brabants. "Mme. d'Albert anticipates all my wants, and makes a spoiled child of me. . . . For M. d'Albert, I love him already as if he were father and brother both. His face is rather haggard-looking, but all the lines and the wavy grey hair indicate the temperament of the artist. I have not heard a word or seen a gesture of his yet that was not perfectly in harmony with an exquisite moral refine-ment—indeed one feels a better person always when he is present."

Three months later, she can still write: "I can say any-thing to M. and Mme. d'Albert. M. d'Albert understands everything, and if Madame does not understand, she believes—that is, she seems always sure that I mean something edifying. She kisses me like a mother, and I am baby enough to find that a great addition to my happiness." It is not surprising, then, that, compared with Geneva, she found "Coventry is a fool to it."

Chapman, in London, was enquiring of the Brays about her translation of Spinoza. To Charles Bray she wrote: "Spinoza and I have been divorced for several months. My want of health has obliged me to renounce all application. I take walks, play on the piano, read Voltaire, talk to my friends, and just take a dose of mathematics every day to prevent my brain from becoming quite soft." She also attended concerts, plays and a course of lectures on Experimental Physics "by M. le Professeur de la Rive, the inventor, among other things, of the electro-plating." Still, Geneva was not quite idyllic. "This terribly severe winter," we find her writing in the February of 1850, "has been a drawback on my recovering my strength. I have lost whole weeks from headache, etc. . . . Decidedly England is the most comfortable country to be in in winter. . . . I hate myself for caring about carpets, easy-chairs, and coal fires—one's soul is under a curse, and can preach no truth while one is in bondage to the flesh in this way; but alas! habit is the purgatory in which we suffer for our past sins."

She returned to England in March, accompanied by d'Albert, and went to stay at Rosehill for a few days before going on to stay with her brother Isaac at Griff and her sister Chrissie at Meriden. Cross, who had a command of meiosis amounting almost to genius, writes: "It will have been seen that she had set her hopes high on the delights of home-coming, and with her too sensitive, impressionable nature, it is not difficult to understand, without attributing blame to anybody, that she was pretty sure to be laying up disappointment for herself."

"Oh the dismal weather, and the dismal country, and the dismal people," she writes to Sara Hennell from Griff. With the death of her father, the mainspring of her day-to-day life had been broken, and now she had to make another life for herself. But how?

It wasn't easy. From Meriden, she returned to Rosehill, to live there for some months. Bessie Parkes, the daughter of Joseph Parkes, of Birmingham, who had helped to finance the publication of her translation of *Leben Jesu*, and in the years to come the mother of Hilaire Belloc, visited her at Rosehill. "Not Abelard in all his glory," Mme. Belloc was to write years later, "not the veritable Isaac Casaubon of French Huguenot fame, not Spinoza in Holland, or Porson in England, seemed to my young imagination more astonishing than this woman, herself not far removed from youth, who knew a bewildering number of learned and modern tongues." Alas, it was one of Marian's off-days: she had one of her headaches and wore "an air of resigned fatigue." But Bessie remembered "her extraordinary quantity of beautiful brown hair."

A more important visitor to Rosehill at this time was Chapman, who was about to buy the *Westminster Review*, the leading organ of philosophical radicalism, which had been in decline ever since John Stuart Mill had given up the editorship ten years earlier. Marian had decided to try to earn her living writing for such reviews, but her only experience of journalism had been writing articles for the *Coventry Herald*. To try her out, Chapman gave her R. W. Mackay's *The Progress of the Intellect as Exemplified in the Religious Development of the Greeks and Hebrews* to review for the *Westminster*. The review she wrote of it was such that she was invited to join Chapman, his family and his publishing firm at 142 Strand, and in November she went.

John Chapman had been a publisher for six years when

Marian joined him. He was two years younger than she and, according to an article on him at the end of the century in the *Athenaeum,* was "so much like the portraits of Lord Byron that among his companions he was always called 'Byron.'" He was born near Nottingham, the son of a druggist, and had been apprenticed to a watchmaker at Worksop, but he had run away from his master and gone to his brother, a medical student in Edinburgh. His brother set him up with a stock of sextants, watches and chronometers and he went to Adelaide, South Australia, as a watchmaker.

He came back to England some years later, saying that he had made a fortune but lost it in a shipwreck. How much of this is true, we don't know. When he married, aged twenty-two, he described himself in the register as "surgeon," and he always claimed that he had studied medicine in Paris; one of his letters refers to the chemist Gay-Lussac as "my old teacher."

If he practised medicine, it cannot have been for long, for in 1843 he wrote a book called: *Human Nature:* A *Philosophical Exposition of the Divine Institution of Reward and Punishment which Obtains in the Physical, Intellectual, and Moral Constitutions of Man: with an introductory Essay. To which is Added a Series of Ethical Observations Written during the Perusal of the Rev. James Martineau's Recent Work entitled "Endeavours after the Christian Life."* He offered this book, which Dr. Haight describes as "little," to a publisher named John Green, who, so far from accepting it, sold Chapman his business. This included, among other things, the British agency of the New England Transcendentalist quarterly, *The Dial,* through which Chapman was able to secure Emerson as one of his authors.

No. 142 Strand was both a publishing house and a boarding house; Chapman had his business on the ground floor, while the rooms above were occupied by Chapman, his family and his lodgers. As a boarding house it was especially popular with visiting American writers; Emerson stayed there for three months in 1848, and other American writers who used it were Greeley, Putnam and Bryant. The evening parties Chapman gave at 142 also made it a centre of London Radical literary life. Chapman had met the Brays through Hennell, whose *Inquiry* he had published. His initial interest in Marian Evans was in her as a possible assistant editor of the *Westminster*, for he must have known that, however great his enthusiasm, he himself lacked the education and intellectual equipment necessary to conduct it.

Chapman was not the sort of man who gets a good press from historians and biographers. He was obviously something of an adventurer, and he was also, it seems clear, an incurable philanderer. When Marian moved into 142 Strand, Chapman was living there with his wife Susanna and his mistress Elisabeth Tilley, who to the public eye was his children's governess. He explained to Marian that he "loved both Susanna and Elisabeth, though each in a different way." It is plain that he was perfectly prepared to love Marian as well and always believed that she loved him. It is likely that she did, for a time. Dr. Haight suggests that it was in Chapman that she found the model for Stephen Guest in *The Mill on the Floss*, and Maggie Tulliver was a permanent side of Marian's character.

The situation in which she found herself was indeed of the old pattern. It is almost as though she wanted to be a member of a harem. The relationship—with a man and with two women—had worked with the Brays; it had been disastrous with the Brabants. It was just as disastrous with the Chapmans. Wife and mistress were alike jealous of her. Chapman helped her to choose a piano for her room, where he listened to her playing Mozart "with much expression"; Susanna thereupon bought a piano for the drawing-room, so that Marian could play more publicly. Chapman then decided to learn German from Marian; Elizabeth also decided to learn the language—from Chapman. The situation, as narrated from day to day in Chapman's diary, is comic enough. Chapman needed them all, Marian as much as the other two, if only because of her importance to him in his business: she had just begun an "Analytical Catalogue" of his publications which would have a value quite additional to its sales potential, since it was a digest of significant religious and philosophical works.

Wife and mistress decided that Chapman and Marian were "completely in love with each other," Susanna having surprised her husband holding Marian's hand. "Bitter remarks" were made; Elisabeth was "all bitterness and icy coldness." The situation was saved temporarily by Marian's going to stay with friends at Highbury; but when she returned, Susanna and Elisabeth joined forces to drive her away. On March 24, 1851, Chapman put her on the train for Coventry. She 'burst into tears" when he explained that he loved Elisabeth and Susanna also, "though in a different way." Marian took with her the material for the Analytical Catalogue.

Six months later Marian was back at No. 142. In the meantime Chapman had been going backwards and forwards between London and Coventry. He had at last concluded his negotiations for the purchase of the *Westminster*. Both as its proprietor and editor, his plans for it were ambitious, and Marian's assistance was essential to them. He had at all costs to square his wife and his mistress to accept Marian, and in the end he prevailed.

He and Marian drew up the prospectus of the new quarterly; it was revised in the light of opinions expressed by leading philosophical radicals, including J. S. Mill. Bray and Chapman arranged the terms and salary on which Marian was to be employed as assistant editor. She was to be in charge of the review section, which consisted of four long articles quarterly on the literature of England, Germany, France and the United States. It was her job to weave each of these into a whole; she herself often wrote parts of the English and American sections, and the French and German regularly. She also helped to gather contributors.

When the first number of the new *Westminster* came out, she could tell the Brays with pride that it was superior both to the *Edinburgh* and the *Quarterly*. It was certainly much more advanced. It was banned from the Edinburgh Select Subscription Library on the score of heresy, from the Sheffield Mechanics' Institute and the libraries of Nottingham.

It was the year of the Great Exhibition. London was full of distinguished foreign visitors, many of whom made their way to No. 142, among them Mazzini, who had contributed to Chapman's *Westminster*. Chapman was always at home on Friday evenings, and there Marian met, among others, political scientists like W. R. Greg; David Brewster, the physicist; James Anthony Froude; Harriet Martineau and Robert Owen. More immediately, though, the most important person to her that she met was Chapman's friend, the philosopher Herbert Spencer, another contributor.

Spencer was almost exactly the same age as Marian and, like her, came from the Midlands. They seem to have become friends on meeting. In a letter to a friend in April, 1852, Spencer speaks of

. . . Miss Evans whom you have heard me mention as the translatress of Strauss and as the most admirable woman, mentally, I ever met. We have been for some time past on very intimate terms. I am very frequently at Chapman's and the greatness of her intellect conjoined with her

womanly qualities and manner, generally keep me by her
side most of the evening.

Spencer was also a journalist, assistant editor of *The
Economist*, and in his *Autobiography* he comments on
this letter:

> For some time before the date of this letter, the occasions of
> meeting her had been multiplied by the opportunities I had
> for taking her to places of amusement. My free admissions
> for two, to the theatres and to the Royal Italian Opera,
> were, during these early months of 1852, much more used
> than they would otherwise have been, because I had fre-
> quently—indeed nearly always—the pleasure of her com-
> panionship in addition to the pleasure afforded by the
> performance.

Marian herself wrote to the Brays on April 22, 1852,
saying that she had been to see *I Martiri* at Covent
Garden with Spencer: "We have agreed that there is no
reason why we should not have as much of each other's
society as we like. He is a good, delightful creature, and
I always feel better for being with him." So the text of
the letter as printed by Cross runs; but collation with
Dr. Haight's edition of George Eliot's letters shows that
Cross had omitted the sentence, "We have agreed that
we are not in love with each other." Was she telling the
truth? A month later, she writes to Sara Hennell:

> My brightest spot, next to my love of *old* friends, is the
> deliciously calm *new* friendship that Herbert Spencer gives
> me. We see each other every day, and have a delightful
> *cameraderie* in everything. But for him my life would be
> desolate indeed. What a wretched lot of old shrivelled crea-
> tures we shall be by-and-by.

Certainly, as we know from Marian's letters to the
Brays, it was assumed on all sides that they were en-
gaged, which was why they could not come down to
Coventry together. In *My Apprenticeship*, Beatrice
Webb, who as a girl knew Spencer, says that it was an

"open secret" that George Eliot was in love with him. But Spencer was not a Chapman. Long before he met Marian, he had his life's course mapped out, and he was not to be deflected from it by emotional relations. He told Beatrice Webb that he had never been in love, and we know from his *Autobiography* that though he often speculated about marriage, he did so in as cold-blooded a manner as possible, drawing up lists of advantages and disadvantages in parallel columns. The disadvantages outweighed the advantages; as he wrote to a friend: "as I see no possibility of being able to marry without being a drudge, why I have pretty well given up the idea." And delightful as he found her company, Marian's attractions were not enough to make him change his mind, no doubt for a reason he stated in his *Autobiography:* "Physical beauty is a *sine qua non* with me; as was once unhappily proved where the intellectual traits and the emotional traits were of the highest." It seems impossible not to believe it was Marian he was thinking of when he wrote that sentence.

The impression one has in the end is that Marian would have married him if he had asked her but that he transferred the question to the high philosophical plane. She had to be content with his company and the knowledge that he had known "but few men" with whom he could "discuss a question in philosophy with more satisfaction" than with her. And obviously his company, even as much as she had of it, was a great deal to her. He was, after all, by a long way the most intellectually distinguished man she had ever known, the only man who can be thought of in any real sense as her equal; and he seems to have seen more in her than anyone else before him had done. He recognised her potentialities:

> It was, I presume, her lack of self-confidence which led her, in those days, to resist my suggestion that she should write novels. I thought I saw in her many, if not all, of

the needful qualities in high degree—quick observation, great powers of analysis, unusual and rapid intuition into others' states of mind, deep and broad sympathies, wit and humour, and wide culture. But she would not listen to any advice. She did not believe she had the required powers.

She needed a deeper relation than Spencer could offer her before belief that she had the required powers was possible to her, a stable and permanent relationship. As it was, her emotional life at this time was anything but stable; as a letter to the Brays in June, 1852, makes clear:

> The opera, Chiswick Flower Show, the French play, and the Lyceum, all in one week, brought their natural consequences of headache and hysterics—all yesterday. At five o'clock I felt quite sure that life was unendurable. This morning, however, the weather and I are both better, having cried ourselves out and used up all our clouds; and I can even contemplate living six months longer.

Her life was almost inconceivably richer than it had been even two years before. She had her own life and was free to move about as she wished. She had new women friends, in particular Bessie Parkes and Barbara Leigh Smith (later Barbara Bodichon), who meant much to her. Her job as assistant editor of the *Westminster* was not only worth while in itself but gave her the entry into some of the most distinguished intellectual circles in the country. She was in the intellectual swim, as few women of her time could be, and she could take the company of eminent men for granted. Her own distinction was increasingly recognised: she was often the only woman at Joseph Parkes's dinners at Savile Row. She was, indeed, almost an honorary man—and that, one guesses, was the trouble, for she was not a man but a woman with a passionate necessity to devote herself to the life of the affections almost to the point of self-immolation.

4

S HE MET GEORGE HENRY LEWES through her work at the
Westminster but came to know him through Spencer, whose close friend he was. At the time, he was
literary editor of the *Leader*, a Radical weekly; indeed,
the first of the English critical weeklies. He had founded
it with his friend Thornton Leigh Hunt, contributing to it
book reviews, dramatic criticism, general articles and
something like a gossip column under the pseudonym
"Vivian." Carlyle called him "the prince of journalists,"
but he was much more than that; he was one of the
most important middle-men of ideas, in the best sense,
of the nineteenth century.

History on the whole has been unfair to him, inevitably perhaps; it is easy to see him merely as the ap-
pendage of George Eliot. But this is not how Marian saw
him. His own creative power—he wrote two novels and
a blank-verse tragedy—was negligible, but he was a
fine critic and, above all, an enterpreneur of ideas. His
Biographical History of Philosophy from Thales to Comte
was a pioneer work in popularisation; his Comte's *Philosophy of the Sciences* was almost the first introduction
of Comte and Positivism to the British public; and, even
more important, an article he wrote in 1843 was the
earliest modern attempt to bring Spinoza to the notice
of the English. Later in life he turned his attention to
biology and wrote *Studies in Animal Life*. And he pro-

duced one work, at any rate, of classic stature, the *Life and Works of Goethe*. His very versatility seems to have made his learning suspect, but he had a wide knowledge of the literature of Greece, Rome, France, Germany, Italy and Spain, of all of whose languages he had at least a good reading knowledge.

When she first met him, Marian hit him off in the phrase "a sort of miniature Mirabeau in appearance." It seems to have been the usual reaction to him. Margaret Fuller described him as "a witty French flippant sort of man." Joan Bennett has suggested that the impression he made on his contemporaries was not unlike that made by Ladislaw on the citizens of Middlemarch. He came of a theatrical family: his father was nicknamed "Dandy" Lewes, and Lewes himself, after studying medicine, had been an actor. He was very ugly; the Carlyles, who liked him, called him "the Ape." Jane Welch Carlyle wrote of him:

> He is the most amusing little fellow in the whole world—if you only look over his unparalleled *impudence* which is not impudence at all but man of genius *bonhomie*. . . . He is the best mimic in the world and full of famous stories, and no spleen or envy, or *bad* thing in him, so see that you receive him with open arms in spite of his immense ugliness.

He was not a gentleman; Frederick Locker-Lampson, who was, says:

> He had long hair, and his dress was an unlovely compromise between morning and evening costume, combining the less pleasing points of both.

Locker-Lampson did not find him "agreeable" but admits that Lewes once made him and Tennyson "laugh heartily by his description of a certain 'noble lord.'"

His social behaviour, too, was often considered outrageous. Eliza Lynn Linton, in her autobiography, *My*

Literary Life in London, remembered him in these terms:

> He was the first audacious man of my acquaintance, and about the most extreme. He had neither shame nor reticence in his choice of subjects, but would discourse on the most delicate matters of physiology with no more perception that he was transgressing the bounds of propriety than if he had been a learned savage. I heard more startling things from Lewes, in full conclave of young and old, married and single, men and women, than I had ever dreamed of or heard hinted at before. And I know that men complained of his after-dinner talk and anecdotes as being beyond the license accorded to, or taken by, even the boldest talkers of the mess-table and the club smoking-room.

In addition to all this, his private life, when Marian met him, was a public scandal. He had married in 1841, and within a year he and his wife had joined with Thornton Leigh Hunt, his wife, his sister and her husband, a sister-in-law and her husband and two unmarried sisters of Mrs. Hunt to take a house in Bayswater, where they lived a communal life. They were, it is clear, very much under the influence of the ideas of Godwin and Shelley. For some years, all was well; the Leweses had three children in the first five years of marriage. And then Mrs. Lewes was, as it were, annexed by Hunt and had two children by him. This Lewes had condoned, perhaps because of the principles of free love he accepted at the time. He showed no resentment and took in his wife's children by Hunt as part of his own family; but, though he continued to live with his wife and the Thorntons, in every real sense his marriage was at an end.

It was, he wrote in his diary some years later, "a very dreary, *wasted* period of my life," the brightest ray in it being his friendship with Herbert Spencer, who did not at the time apparently know the circumstances of Lewes's domestic life but wrote later:

. . . alike then and afterwards I was impressed by his for-
giving temper and generosity. Whatever else may be
thought, it is undeniable that he discharged the responsibilities which devolved upon him with great conscientious-
ness, and at much cost in self-sacrifice, notwithstanding
circumstances which many men would have made a plea
for repudiating them.

Such was the man who, one afternoon in 1853, called
on Marian with Spencer and, when Spencer left, stayed
behind. On the face of it—in the public view, at any rate
—they would seem to be so far apart temperamentally as
scarcely to have points of contact, the "witty French
flippant sort of man" and the George Eliot of the novels
who so often suggests Wordsworth's personification of
Duty, "Stern daughter of the voice of God." And then one
remembers how Dorothea Brooke and Will Landislaw
responded to each other; and, in fact, as one sees from
her letters, Marian during the months before had been
speaking more and more warmly of Lewes. "We had a
pleasant evening last Wednesday," she had written to
Sara Hennell in March. "Lewes, as always, genial and
amusing. He has quite won my liking, in spite of myself";
and, a month later, to Mrs. Bray: "People are very good
to me. Mr. Lewes is especially kind and attentive, and
has quite won my regard, after having had a good deal
of my vituperation. Like a few other people in the world,
he is much better than he seems. A man of heart and
conscience wearing a mask of flippancy."

The inference is that when he stayed behind on that
afternoon in 1853, after Spencer had left, he poured his
heart out to her. Whether he then proposed marriage—
or its equivalent—we do not know. But the very situation,
the appeal from the man to the woman, was one to
which, knowing Marian in her life and through the
personae of herself she presents in her fiction (both
Maggie and Dorothea), one feels she could not help

responding. And the very desperateness of the situation, for Lewes's condonation of his wife's adultery made divorce impossible, can only have heightened the appeal for Marian. She moved from 142 Strand into lodgings in Hyde Park Square. When Lewes was ill she corrected his proofs and did his work for him. She arranged with Chapman to continue on the *Westminster* until the following April. On July 20, 1854, she left with Lewes for Germany, where they stayed for eight months in Weimar and Berlin. In her eyes they were man and wife.

She made a statement of her principles fourteen months later, in a letter to Mrs. Bray, which shows also how even her oldest friends had misinterpreted her conduct:

If there is any one action or relation of my life which is and always has been profoundly serious, it is my relation to Mr. Lewes. It is, however, natural enough that you should mistake me in many ways, for not only are you unacquainted with Mr. Lewes's real character and the course of his actions, but it is also several years now since you and I were much together, and it is possible that the modifications my mind has undergone may be quite in the opposite direction of what you imagine. No one can be better aware than yourself that it is possible for two people to hold different opinions on momentous subjects with equal sincerity, and an equally earnest conviction that their respective opinions are alone the truly moral ones. If we differ on the subject of the marriage laws, I at least can believe of you that you cleave to what you believe to be good; and I don't know of anything in the nature of your views that should prevent you from believing the same of me. *How far* we differ, I think we neither of us know, for I am ignorant of your precise views; and apparently you attribute to me both feelings and opinions which are not mine. We cannot set each other quite right in this matter in letters, but one thing I can tell you in a few words. Light and easily broken ties are what I neither desire theoretically nor could live for practically. Women who are satisfied with such ties do *not* act as I have done. That any unworldly, unsuperstitious person who is sufficiently acquainted with the realities of life can pronounce my relation to Mr. Lewes immoral, I can only understand by remembering how subtle and complex are the influences that mould opinion. But I *do* remember this: and I indulge in no arrogant or uncharitable thoughts about those who condemn us, even though we might have expected a somewhat different verdict. From the majority of persons, of course, we never looked for anything but condemnation. We are leading no life of self-indulgence, except indeed that, being happy in each other, we find everything easy. We are working hard to provide for others better than we provide for ourselves, and to fulfil every responsibility that lies upon us. Levity and pride would not be a sufficient basis for that. Pardon me if, in vindicating myself from some unjust conclusions, I seem too cold and self-asserting. I should not care to vindicate myself if I did not love you and desire to relieve you of the pain which you say these conclusions have given you. Whatever I may have misinterpreted before, I do not misinterpret your letter this morning, but read in it nothing else than love and kindness towards me, to which my heart fully answers yes. I should like never to write about myself again; it is not healthy to dwell on one's own feelings and conduct, but only to try and live more faithfully and lovingly every fresh day. I think not one of the endless words and deeds of kindness and forbearance you have ever shown me has vanished from my memory. I recall them often, and feel, as about everything else in the past, how deficient I have been in almost every relation of my life. But that deficiency is irrevocable, and I can find no strength or comfort except in "pressing forward towards the things that are before," and trying to make the present better than the past. . . .

Her position when they returned to England was, as that letter shows, anything but easy. At first her friends wrote to her as Miss Evans, while landladies knew her as

Mrs. Lewes; until she found it necessary to insist, at the price of friendship, that she be called Mrs. Lewes. In her own mind, she was clear that morally she was Lewes's wife, and his three sons accepted her as such, calling her mother. But even in 1860, when she was famous as George Eliot, we find Jane Welch Carlyle saying:

> When one was first told that the strong woman of the *Westminster Review* had gone off with a man we all knew, it was as startling an announcement as if one had heard that a woman of your acquaintance had gone off with the strong man at Astley's. . . . That the partners should set up as moralists was a graver surprise. To renounce George Sand as a teacher of morals was right enough, but it was scarcely consistent with making so much of our George in that capacity. A marvellous teacher of morals, surely, and still more marvellous in that other character, for which nature has not provided her with the outfit supposed to be essential.

Mrs. Carlyle may well have been piqued that, ignorant of who George Eliot was, she had written her a fan letter about *Adam Bede;* but, as Eliza Lynn Linton wrote years later:

> Society was at first as stern to George Eliot after domestic intimacy with Lewes as Mrs. Carlyle had been. I remember hearing an instance of this some years after the connection was formed. Lewes and George Eliot once thought of establishing a domicile in Kent and a south-eastern, semi-suburb of London much tenanted by wealthy city people was chosen. When news of the intention reached the denizens of the region a council of male and female heads of families was held to consider whether George Eliot should be "received." It was decided that she should not.

In the end, she prevailed—or almost so. As long as she lived with Lewes, her brother and sister would have no communication with her; and even in 1877, when she made her triumphal visit to Cambridge and so impressed

F. W. H. Myers, she was not allowed to enter Girton College, the new foundation for women students, although she had subscribed towards it. In the last analysis, that she did prevail is due to the fact that, becoming Mrs. Lewes, she also became George Eliot; and for that, Lewes's is the credit.

"September 1856 made a new era in my life," she recalled in her journal, "for it was then I began to write fiction." It had always been a vague dream of hers that one day she might write a novel, though, as we have seen, Spencer had been unable to bring her to the point of translating it into reality. She had, apparently, written a first chapter "describing a Staffordshire village and the life in the neighbouring farm-houses," but as the years passed had lost "any hope that I should ever be able to write a novel, just as I desponded about everything else in my future life. I always thought I was deficient in dramatic power, both of construction and dialogue."

This first chapter was among the papers she took with her to Germany, and one evening in Berlin she read it to Lewes. It suggested to him the possibility of her being able to write a novel, "though he distrusted—indeed disbelieved in—my possession of any dramatic power."

> Still, he began to think that I might as well try sometime what I could do in fiction; and by-and-by, when we came back to England, and I had greater success than he ever expected in other kinds of writing, his impression that it was worth while to see how far my mental power would go, towards the production of a novel, was strengthened. He began to say very positively, "You must try and write a story," and when we were at Tenby he urged me to begin at once. . . . One morning as I was thinking what should be the subject of my first story, my thoughts merged themselves into a dreamy doze, and I imagined myself writing a story, of which the title was "The Sad Fortunes

of the Reverend Amos Barton." I was soon wide awake again and told G. He said, "Oh, what a capital title!" and from that time I had settled in my mind that this should be my first story. George used to say, "It may be a failure —it may be that you are unable to write fiction. Or perhaps it may be just good enough to warrant your trying again." Again, "You may write a *chef d'oeuvre* at once—there's no telling." But his prevalent impression was, that though I could hardly write a *poor* novel, my effort would want the highest quality of fiction—dramatic presentation. He used to say, "You have wit, description, and philosophy—those go a good way towards the production of a novel. It is worth while for you to try the experiment.". . .

But when we returned to Richmond, I had to write my article on Silly Novels, and my review of Contemporary Literature for the "Westminster," so that I did not begin my story till September 22. After I had begun it, as we were walking in the park, I mentioned to G. that I had thought of the plan of writing a series of stories drawn from my own observation of the clergy, and calling them "Scenes from Clerical Life," opening with "Amos Barton." He at once accepted the notion as a good one—fresh and striking; and about a week later, when I had read him the first part of "Amos," he had no longer any doubt about my ability to carry out the plan. The scene at Cross Farm, he said, satisfied him that I had the very element he had been doubtful about—it was clear I could write good dialogue. There still remained the question whether I could command any pathos; and that was to be decided by the mode in which I treated Milly's death. One night G. went to town on purpose to leave me a quiet evening for writing it. I wrote the chapter from the news brought by the shepherd to Mrs. Hackit, to the moment when Amos is dragged from the bedside, and I read it to G. when he came home. We both cried over it, and then he came up to me and kissed me, saying, "I think your pathos is better than your fun."

Lewes sent *Amos Barton* to Blackwood as the work of a friend, whom Blackwood assumed was a clergyman.

Marian took the name of George Eliot when she had to reply to a letter from Blackwood; according to Cross, because "Eliot was a good mouth-filling, easily-pronounced word" and George was Lewes's Christian name. The three stories that make up *Scenes of Clerical Life* were serialised in *Blackwood's* and then published by him in two volumes.

They were received enthusiastically. Dickens, in a letter to George Eliot, more or less accused her of being a woman; Thackeray believed she was not. At this time, the only person in the secret apart from the Blackwoods was Spencer. Lewes did his best to keep the identity of George Eliot a secret, even to the point of lying to Chapman; but after a Mr. Liggins, a dissenting clergyman of Nuneaton, had been credited with being George Eliot, it was made public.

5

THE PUBLICATION OF *Adam Bede,* in 1859, established Marian as the leading woman novelist of the day and, indeed, as second only to Dickens and Thackeray generally. From then on, Marian Evans was George Eliot, the great Victorian, the centre of an admiring and often adoring circle, to whose Sunday afternoons at home at Blandford Square or, later, Regent's Park it was a privilege and sometimes an ordeal ("I never read criticisms of my own work," she told a young reviewer who mentioned his ardent notice of *Daniel Deronda*) to attend.

One of those who attended regularly was Oscar Browning. He has left a memorable account of the at-homes in his *Life of George Eliot:*

> It was a double drawing-room without folding doors, decorated by Owen Jones, and hung with Leighton's drawings for the illustration of *Romola.* A bow window, with casements down to the ground, looked on to the garden. Mrs. Lewes generally sat in an armchair at the left of the fireplace. Lewes generally stood or moved about in the back drawing-room, at the end of which was the grand piano, on which, so far as I am aware, she never played during these receptions. In the early days of my acquaintance the company was small, containing more men than women. Herbert Spencer and Professor Beesley were constant visitors. The guests closed round the fire and conversation was general. At a later period the company increased, and those who

wished to converse with the great authoress whom they
had come to visit took their seats in turn at the chair by
her side. She always gave us of her best. Her conversation
was deeply sympathetic, but grave and solemn, illumined
by happy phrases and thrilling tenderness, but not by
humour. Although her features were heavy, and not well-
proportioned, all was forgotten when that majestic head
bent slowly down, and the eyes were lit up with a penetrat-
ing and lively gaze. She appeared much greater than her
books. Her ability seemed to shrink beside her moral
grandeur. She was not only the cleverest, but the best
woman you had met. You never dared to speak to her of
her works; her personality was so much more impressive
than its product. At a later time the string of visitors
became fatiguing to those who remembered the old days.
The drawing-room was enlarged to hold them; and three
fashionably dressed ladies, sweeping in, occupied the sofa,
and seemed to fill the room. These Sunday afternoon recep-
tions were a great strain upon her strength. When the last
visitor had departed she would, if the weather were fine,
seek refreshment in a brisk walk to dispel her headache,
and to call back the circulation into her feet, the icy
coldness of which was one of her perpetual trials.

That was George Eliot the great novelist who was
more than a mere novelist—George Eliot the Sibyl. The
house at North Bank, Regent's Park, was called The
Priory, and Lewes often spoke of their Sunday afternoons
facetiously as "religious services at the Priory" and of
George Eliot as "Madonna." She was indeed the centre
of something very much like a cult, of which Lewes
was the devoted but also, one feels, slightly amused high
priest: in their private life she was his "Polly."

One gets a fascinating glimpse of her in the autobiography, which was never intended for publication, of
the journalist and social worker Edith Simcox, who was
without question lesbian. She adored George Eliot with
an adoration at once sexual and quasi-religious; she

would, given the opportunity, which was sometimes snatched rather than given, kiss her feet. George Eliot did not respond; but though she was embarrassed she was also touched by Edith's devotion, and plainly she understood it. Edith called her "Mother," and though she did not approve in this instance, George Eliot, who as a little girl at school was nicknamed "Little Mamma," by this time was called "Mother" by more than one young woman and signed her letters to them as such.

What is especially interesting is Lewes's attitude in all this. Edith Simcox was perfectly at ease with him; plainly he understood, too, and seems to have treated her as fellow-conspirator in adoration. He made what could have been a difficult relation cosy. That indeed was one of the functions he took upon himself. He had in a very real sense, which does not detract at all from her own genius, made George Eliot, because it was through him that she discovered it; and having made her, he cherished and protected her, shielding her from hostile reviews, guarding her sensitivity from the pinpricks of the world. She, for her part, gave him much, not the least, paradoxically, being, through her insistence that their liaison was a true marriage and must be accepted as such by all who wished to know her, a respectability in the eyes of the world that without her he would probably never have achieved.

The seal was set on it in May, 1877, when they dined in a company that included Queen Victoria's daughter, the Princess Louise, who insisted on being presented to George Eliot instead of George Eliot's being presented to her. If further endorsement of their respectability was required, it came a year later when, at a dinner party which included among the guests the Dean of Westminster, the Bishop of Peterborough and Lord and Lady Ripon, they dined with the Crown Prince and Princess of Germany, the Princess being another of Queen Victoria's daughters. George Eliot described the encounter in a letter to Mrs. Bray:

> The royalties did themselves much credit. The Crown Prince is really a grand-looking man, whose name you would ask for with expectation, if you imagined him no royalty. He is like a grand antique bust—cordial and simple in manners withal, shaking hands, and insisting that I should let him know when we next come to Berlin, just as if he had been a Professor Gruppe, living *au troisième. She* is equally good-natured and unpretending, liking best to talk of nursing soldiers, and of what her father's taste was in literature. She opened the talk by saying, "You know my sister Louise"—just as any other slightly embarrassed mortal might have done. . . .
> We go to Oxford tomorrow (to the Master of Balliol).

The Leweses were not only eminently respectable—their very presence now conferred respectability on any company they were in. Yet it is clear that George Eliot still felt painfully the vulnerability of her position. This comes out plainly in an incident reported by Oscar Browning. In October, 1879, the Leweses went to stay with friends in Cambridgeshire. One of the other guests was Turgenev, of whom Cross says, "I remember George Eliot telling me that she had never met any literary man whose society she enjoyed so thoroughly and so unrestrainedly. . . . They had unnumerable bonds of sympathy." Turgenev told her of a play he had seen in Paris about a woman who had been deserted with her young children by her husband and then lived for twenty years with a man who had brought up the children as his own. The real father returned and revealed himself to his eldest son; when the man he had always believed was his father came home that evening and kissed the daughter in his usual way, the son slapped his face, exclaiming, "You have not the right to do that," to the great applause of the audience. Turgenev alone stood up in his box and hissed.

George Eliot made Turgenev repeat the story to the whole party, who were naturally fascinated with its relevance to the emotional lives of both novelists. At dinner that night, Lewes proposed Turgenev's health as the greatest living novelist; he refused to accept it, saying in broken English that the title was George Eliot's.

Since 1876 the Leweses had been living at Witley, in Surrey, where they had bought an architectural monstrosity of a house which looked like the issue of an unnatural union between an Elizabethan mansion and a Swiss chalet. There they lived a life of seclusion, though they had agreeable neighbours and were in visiting distance of such friends as the Tennysons and Du Mauriers. Lewes was increasingly ill, but, as Cross recalls: "Even on his worst days he had always a good story to tell; and I remember on one occasion, in the drawing-room at Witley, between two bouts of pain, he sang through, with great *brio,* though without much voice, the greater portion of the tenor part in the *Barber of Seville*—George Eliot playing his accompaniment, and both of them thoroughly enjoying the fun."

The visit they made to meet Turgenev was the last journey they made together. Of Lewes, Cross writes: "Nothing but death could quench that bright flame": it did so on November 28, 1878.

For many weeks George Eliot saw no one except Lewes's son Charles and the very few people she was obliged to receive on essential business. On January 1, 1879, she wrote in her diary—it was the only entry— "Here I and sorrow sit." A week later she wrote to a friend that she was "a bruised creature," shrinking "even from the tenderest touch." She busied herself with the foundation in Lewes's name of a studentship in physiology at Cambridge, the editing of manuscripts he had

left behind and the preparation for the press of her
essays, *Impressions of Theophrastus Such.*

Madame Bodichon saw her early in June and wrote to
a friend: "She was more delightful than I can say, and
left me in good spirits for her—though she is wretchedly
thin, and looks in her long, loose, black dress like the
black shadow of herself. She said she had so much to do
that she must keep well—'the world was so *intensely
interesting.*' . . . We both agreed in the great love we had
for life. In fact, I think she will do more for us than ever."

She had, in fact, been seeing John Walter Cross "constantly," according to his own account, since April 22,
when she had written to him: "I am in dreadful need of
your counsel. Pray come to me when you can—morning,
afternoon, or evening." He came; and later he wrote:

My mother had died in the beginning of the previous
December—a week after Mr. Lewes; and as my life had
been very much bound up with hers, I was trying to find
some fresh interest in taking up a new pursuit. Knowing
very little Italian, I began Dante's "Inferno" with Carlyle's
translation. The first time I saw George Eliot afterwards,
she asked me what I was doing, and, when I told her,
exclaimed, "Oh, I must read that with you." And so it was.
In the following twelve months we read through the "Inferno" and the "Purgatorio" together—not in a *dilettante*
way, but with minute and careful examination of the construction of every sentence. The prodigious stimulus of
such a teacher *(cotanto maestro)* made the reading a real
labour of love. Her sympathetic delight in stimulating my
newly awakened enthusiasm for Dante, did something to
distract her mind from sorrowful memories. The divine poet
took us into a new world. It was a renovation of life. At
the end of May I induced her to play on the piano at
Witley for the first time; and she played regularly after
that whenever I was there, which was generally once or
twice a week. . . .

Besides Dante, we read at this time a great many of

Sainte-Beuve's "Causeries" and much of Chaucer, Shakespeare, and Wordsworth.

The reading of Dante together proved to be a labour of love in a double sense. Cross was a wealthy business man, more than twenty years younger than George Eliot. They had first met in Rome in 1869, when, as he said himself, "I was better acquainted with George Eliot's books than with any other literature." They met again four months later at Weybridge, where Cross's mother lived. "A day," Cross wrote, "did the work of years. Our visitors had come to the house as acquaintances, they left as lifelong friends." It was natural that she should turn to him for advice after Lewes's death and that he should turn to her after the death of his mother. A "bond of mutual dependence"—Cross's words—was formed between them. When, towards the end of 1879, George Eliot began to move in society again, she formed the habit of going frequently to the National Gallery and other art exhibitions with Cross. "This constant association," Cross wrote, "engrossed me completely, and was a new interest to her." On April 9, 1880, they decided to marry, "as privately as might be found practicable," for, as George Eliot wrote to Cross's sister, "I quail a little in facing what has to be gone through—the hurting of many whom I care for."

But it was, as she said, a "wonderful renewal of my life." In May she wrote to Barbara Bodichon:

I am going to do what not very long ago I should myself have pronounced impossible for me, and therefore I should not wonder at any one else who found my action incomprehensible. By the time you receive this letter I shall (so far as the future can be a matter of assertion) have been married to Mr. J. W. Cross, who, you know, is a friend of years, a friend much loved and trusted by Mr. Lewes, and who, now that I am alone, sees his happiness in the dedication of his life to me.

They were married the day after she wrote that letter, on May 6, at St. George's, Hanover Square, and left immediately for France, Switzerland and Italy.

She was right in thinking that many would find her action incomprehensible. There was the discrepancy in age between Cross and herself. Her Positivist friends were outraged because she had broken Comte's laws against second marriages. Eliza Lynn Linton, believing that George Eliot had succumbed to a craving for legal marriage, called it "the crowning act of weakness." Others were disconcerted at the very haste of the marriage, within eighteen months of Lewes's death; it seemed to contradict the constancy and the mutual devotion for which she and Lewes had become almost synonymous. Her brother Isaac, on the other hand, who had refused to communicate with her personally throughout her life with Lewes, wrote and congratulated her.

The Crosses returned from their honeymoon at the end of July. Their life together was brief, for on December 22, 1880, she died, after catching a chill at a concert in the St. James's Hall. She was buried in the grave next to Lewes in Highgate Cemetery. Cross survived her forty-four years and left behind him not only *George Eliot's Life* but *Impressions of Dante and of the New World, with a Few Words in Bimetallism.*

❖《❖《❖《❖《❖ GEORGE ELIOT ❖》❖》❖》❖》❖

The Novels

1

G EORGE ELIOT has been called the first modern English
novelist. The truth of the generalisation depends
upon the angle from which it is approached. Some light
on it, however, may be thrown by a roll call of the
novels that appeared in Europe within four or five years
either side of the publication of *Scenes of Clerical Life*
(1858) and *Adam Bede* (1859). Turgenev's *Sportsman's
Sketches* appeared in 1852 and *Rudin* in 1856, Tolstoy's
Sevastopol and Flaubert's *Madame Bovary* in 1855,
Dostoevsky's *The House of the Dead* in 1861, and Turgenev's *Fathers and Sons* in 1863. None of these novels
had the slightest influence upon her; indeed, so far as one
can tell, Turgenev is the only novelist among them that
she read. Nevertheless, it is in the context of the European novels of her time that we read and think of her
today rather than in that of her purely English contemporaries Dickens, Thackeray and Trollope.

Yet, seen formally as a novelist, it is with them that she
belongs. Technically, she is the novelist as omniscient
narrator, relating an action long after its conclusion as
though she were God, commenting on the action, telling
us which characters in it are to be admired, which to be
deplored and why. It is one way of writing novels and,
as the best examples show, a good one. It depends for
complete artistic success, however, upon the novelist's
tone of voice, for we are in this kind of novel inescapably

in the presence of the story-teller telling his story. The story-teller, indeed, is almost as important as the story he tells, for it is the sense we have of him through the manner and style of his narration that mediates between us and the story he tells. He establishes the values of his story as much by his way of saying it as by what he says. In this kind of novel, we have, as it were, one foot in the personal essay, which is why the author's personal manner of speaking is so important; "writing," as Sterne said, "when properly managed, is but a different name for conversation."

In English, the great master of this kind of novel is Fielding, with Thackeray, at a lower level of seriousness, following him. Fielding, in particular, has a colloquial ease of style, at once civilised and eloquent, that suggests a man of the world—but a man of the world who is something more than the phrase itself would indicate—talking familiarly with his equals. His lightness of touch does not for one moment invalidate the tremendous seriousness of the implications of what he is telling us, for the manner itself is, we feel, the expression of a temperament allied to a vast experience of life, the experience, in fact, of an unillusioned and unshockable yet deeply compassionate police-court magistrate who understands himself at least as well as the felons he meets in the dock each day.

In the last analysis, tone is a function of style; and in this regard, it is evident, in terms of the kind of novel she was writing, that George Eliot was lacking. Compared with Fielding's and Thackeray's, the authorial comment in which her stories are embodied is intrusive, indeed obtrusive. Fielding persuades us of the truth of his interpretation of what he is narrating by his appeal both to sweet reasonableness and to worldly experience; George Eliot, in contrast, lectures us and sometimes even hectors us. She lacks tact, as she lacks wit, except a ponderous irony. She gives the impression, in fact, of not quite knowing whom she is addressing. Her style is not subtle enough or easy enough; she is too self-conscious, too anxious that we should not misunderstand the point that the incidents and episodes of her novels should make themselves.

One may speculate why this should have been so. She seldom, as her letters show, when writing in her own person, wrote a sentence that gives pleasure in itself. She is, somehow, too much in the way, unable, perhaps, to distinguish between seriousness and solemnity. Over-earnest herself, she assumes her readers are over-earnest, and she has great faith in polysyllables in their own right. Admittedly, the convention in which she was writing was breaking down even while she was working in it, and she may have been half aware of this. But much of the strain seems to come from the large ends she proposed to herself in the writing of fiction.

It is these, of course, that link her with her great contemporaries in the European novel and to some extent remove her from the company of her English coevals. She would have found, for example, Wilkie Collins's formula for the writing of fiction, "Make 'em laugh, make 'em cry, make 'em wait," which is behind a great deal of Dickens's work also, quite unworthy of emulation; and though she might have accepted Trollope's description of the novel—". . . a picture of common life enlivened by humour and sweetened by pathos. To make this picture worthy of attention, the canvas should be crowded with real portraits, not of individuals known to the world or to the author, but of created personages impregnated with traits of character which are known"—she would have wanted to add much more. Indeed, she did so.

In a letter written in 1861, on an article on *The Mill on the Floss*, she writes: "It is a comfort to me to read any criticism which recognises the high responsibilities of literature that undertakes to represent life. The ordinary tone about art is that the artist may do what he will, provided he pleases the public." And "the high responsibilities of literature that undertakes to represent life" she does, in fact, enlarge on in her early fiction—precisely in some of those authorial asides and comments which now seem flaws in her work.

Right at the beginning of her career, she expressed one of her chief aims in fiction in these words:

> The Rev. Amos Barton . . . was, you perceive, in no respect an ideal or exceptional character; and perhaps I am doing a bold thing to bespeak your sympathy on behalf of a man who was so very far from remarkable. . . . "An utterly uninteresting character!" I think I hear a lady reader exclaim—Mrs. Farthingale, for example, who prefers the ideal in fiction; to whom tragedy means ermine tippets, adultery, and murder; and comedy, the adventures of some personage who is quite a "character."
>
> But, my dear madam, it is so very large a majority of your fellow-countrymen that are of this insignificant stamp. At least eighty out of a hundred of your adult male fellow-Britons returned in the last census are neither extraordinarily silly, nor extraordinarily wicked, nor extraordinarily wise; their eyes are neither deep and liquid with sentiment, nor sparkling with suppressed witticisms; they have probably had no hairbreadth escapes or thrilling adventures; their brains are certainly not pregnant with genius, and their passions have not manifested themselves at all after the fashion of a volcano. They are simply men of complexions more or less muddy, whose conversation is more or less bald and disjointed. Yet these commonplace people—many of them—bear a conscience, and have felt the sublime prompting to do the painful right; they have felt their unspoken sorrows, and their sacred joys; their hearts have perhaps gone out towards their first-born, and they mourned over the irreclaimable dead. Nay, is there not a pathos in their very insignificance—in our comparison of their dim and narrow existence with the glorious possibilities of that human nature which they share?

Depend upon it, you would gain unspeakably if you would learn with me to see some of the poetry and the pathos, the tragedy and the comedy, lying in the experience of a human soul that looks out through dull grey eyes, and that speaks in a voice of quite ordinary tones. . . .

"Depend upon it, you would gain unspeakably if you would learn with me . . .": the whole passage is a perfect example of authorial intervention at its most inept; the tone is disastrous in the relentlessness of its nagging. But that is not the point here. It is a clear statement of George Eliot's aim of enlarging human sympathies through fiction.

In *Adam Bede,* again speaking in her own person as the author, she enlarges on this and at the same time quite clearly states her own attitude to her material, her conception of her duties as a novelist. It occurs in the first chapter of the second book, a chapter entitled ominously "In Which the Story Pauses a Little." It begins:

"This Rector of Broxton is little better than a pagan!" I hear one of my readers exclaim. "How much more edifying it would have been if you had made him give Arthur some truly spiritual advice. You might have put into his mouth the most beautiful things—quite as good as reading a sermon."

Certainly I could, if I held it the highest vocation of the novelist to represent things as they never have been and never will be. Then, of course, I might refashion life and character entirely after my own liking; I might select the most unexceptionable type of clergyman, and put my own admirable opinions into his mouth on all occasions. But it happens, on the contrary, that my strongest effort is to avoid any such arbitrary picture, and to give a faithful account of men and things as they have mirrored themselves in my mind. The mirror is doubtless defective; the outlines will sometimes be disturbed, the reflection faint or confused; but I feel as much bound to tell you as precisely as I can what that reflection is, as if I were in the witness-box narrating my experience on oath.

One may set side by side with that a passage from Henry James's essay *The Art of Fiction,* written twenty-five years later:

> Certain accomplished novelists have a habit of giving themselves away which must often bring tears to the eyes of people who take their fiction seriously. I was lately struck, in reading over many pages of Anthony Trollope, with his want of discretion in this particular. In a digression, a parenthesis or an aside, he concedes to the reader that he and this trusting friend are only "making believe." He admits that the events he narrates have not really happened, and that he can give his narrative any turn the reader may like best. Such a betrayal of a sacred office seems to me, I confess, a terrible crime, and it shocks me every whit as much in Trollope as it would have shocked me in Gibbon or Macaulay.

The reference to Gibbon and Macaulay shows where the appeal lies; the "sacred office" is that of the novelist as historian. Trollope's practice, James asserts, "implies that the novelist is less occupied in looking for the truth than the historian, and in doing so it deprives him at a stroke of all his standing-room." In her own way, George Eliot is saying something very similar: the novelist is on oath to tell the truth. It is her determination here that removes her from the company of Trollope, excellent novelist as he was, into the much more serious company of her European contemporaries.

George Eliot goes on, in this chapter in *Adam Bede,* to describe the kind of art she is attempting:

> So I am content to tell my simple story, without trying to make things seem better than they are; dreading nothing, indeed, but falsity, which, in spite of one's best efforts, there is reason to dread. Falsehood is so easy, truth so

difficult. . . . Examine your words well, and you will find
that even when you have no motive to be false, it is a very
hard thing to say the exact truth, even about your own
immediate feelings—much harder than to say something fine
about them which is *not* the exact truth.

It is for this rare, precious quality of truthfulness that I
delight in so many Dutch paintings, which lofty-minded
people despise. I find a source of delicious sympathy in
these faithful pictures of a monotonous homely existence,
which has been the fate of so many more among my fellow-
mortals than a life of pomp or of absolute indigence, of
tragic suffering or of world-stirring actions. I turn, without
shrinking, from cloud-borne angels, from prophets, sibyls,
and heroic warriors, to an old woman bending over her
flower-pot, or eating her solitary dinner, while the noonday
light, softened perhaps by a screen of leaves, falls on her
mob-cap, and just touches the rim of her spinning wheel,
and stone jug, and all those cheap common things which
are the precious necessaries of life to her. . . .

Again:

In this world there are so many of these common coarse
people, who have no picturesque sentimental wretchedness!
It is so needful that we should remember their existence,
else we may happen to leave them quite out of our religion
and philosophy, and frame lofty theories which only fit a
world of extremes. Therefore let Art always remind us of
them; therefore let us always have men ready to give the
loving pains of a life to the faithful representing of com-
monplace things—men who see beauty in these common-
place things, and delight in showing how kindly the light
of heaven falls on them. There are few prophets in the
world; few sublimely beautiful women; few heroes. I can't
afford to give all my love and reverence to such rarities:
I want a great deal of those feelings for my everyday
fellow-men, especially for the few in the foreground of the
great multitude, whose faces I know, whose hands I touch,
for whom I have to make way with kindly courtesy. . . . It
is more needful that I should have a fibre of sympathy

connecting me with that vulgar citizen who weighs out
my sugar in a vilely-assorted cravat and waistcoat, than
with the handsomest rascal in red scarf and green feathers;
—more needful that my heart should swell with loving
admiration at some trait of gentle goodness in the faulty
people who sit at the same hearth with me, or in the
clergyman of my own parish, who is perhaps rather too
corpulent, and in other respects is not an Oberlin or Tillot-
son, than at the deeds of heroes whom I shall never know
except by hearsay, or at the sublimest abstract of all clerical
graces that was ever conceived by an able novelist.

In passages of theory like these, as generally in her
practise, George Eliot appears as the quintessential
novelist, the sign of whom, from Cervantes, Defoe and
Fielding onwards, has always been a deep suspicion
of fiction in the sense of the false and contrived.

George Eliot, then, inherited the convention of the
omniscient narrator telling his story in his own voice and
allowing himself whatever comments, digressions and
asides he cares to make. She remained faithful to it until
the end; but, in fact, after *The Mill on the Floss*, her
digressions were much more curbed, her asides much less
flagrant and her comment much more closely woven
into the texture of her narrative. After *Adam Bede* there
is nothing quite so obtrusive as the paragraphs quoted
above, and when one turns from *The Mill on the Floss* to
Felix Holt, written seven years later, one is conscious
immediately of a sharp increase in dramatisation; the
action is allowed much more to speak for itself. W. J.
Harvey, in *The Art of George Eliot*, gives the ratio of
omniscient intrusions in terms of pages as, in *Adam Bede*,
1:10; in *The Mill on the Floss*, 1:14; in *Middlemarch*,
1:33. Once she had learnt how to control them, they are,
of course, much more acceptable, partly because of the
very amplitude of her novels and the leisurely pace at
which they are narrated; for her fiction conforms essen-

tially to Smollett's definition of the novel as "a large diffused picture, comprehending the characters of life, disposed in different groups and exhibited in various attitudes, for the purpose of a uniform plan."

Within the convention she followed, she was an innovator in the high seriousness which she claimed for herself as a novelist. In this respect she may be seen as the watershed between the traditional novel and the modern; after her come Hardy, James and Lawrence, who were all in one way or another indebted to her. But she was an innovator in another sense also: in her insistence upon the evolutionary forces that shape society. With the exceptions of *Romola*, a historical novel, and *Daniel Deronda*, all her fiction is set in a past that is not, indeed, remote but that, as it were, is truly over. The period of time spanned is that from the last decade of the eighteenth century to about the time of the first Reform Act, 1832, the period, in other words, of her own childhood and the twenty years before her birth. It is seen always from the vantage-point of the mid-Victorian age and slightly later in which she was writing, and by contrast with those turbulent years of rapid progress, it appears still a traditional society. Time and again, George Eliot makes great play with the contrast, the outstanding example being the prologue to *Felix Holt*. It appears the more traditional because it is a provincial society, far removed from the centre of things; even in *Middlemarch*, which is set partly in Coventry only a few years before she was living there herself, there are no characters equivalent to the Brays and the Hennells, for Lydgate, after all, is an outsider and Ladislaw a bird of passage.

To this provincial society—and it is only in *Middlemarch* that a substantial town is at its centre—George Eliot looks back with considerable nostalgia. She does so because it *is* a traditional society, in which the rhythms of life are established and duties are known. But—and this is the important thing—it is not a static society; it is one in process of being transformed by forces outside it. It would be wrong to call them revolutionary; rather, they are evolutionary. This indeed is the point of view from which George Eliot is writing.

In all these novels, the picture of society is built up slowly and leisurely, through figures who can be considered representative and whose backgrounds are presented in a sober, minute realism that can be equated with that of the seventeenth-century Dutch paintings she so much admired. She is describing, in other words, society at a moment of transition, when new evolutionary factors in society are coming into play. There is the incursion into a remote village of Methodism in *Adam Bede;* the consequences of Radical politics and of political controversy in *Felix Holt.* The disruptive, evolutionary forces that shape *Middlemarch* include not only the politics of the great debate that culminated in the first Reform Act but also the coming of the railway and new developments in medicine and public health. The world she portrays is one riven by great historical events, and this is so even when, as in *Adam Bede* and *Silas Marner,* it is a world quite marginal to the great world; remote as they are from the scene of the earthquake, Hayslope and Raveloe cannot escape the tremors it sets up.

There is another way in which George Eliot established the fact of change and evolution in the society she is describing. Her characters are not all of one piece. Her nostalgia is for the deeply rooted in traditional ways of life, for the Poysers and the Reverend Mr. Irwine. in *Adam Bede,* for the Dodson sisters in *The Mill on the Floss,* for the Garths in *Middlemarch.* It is none the less nostalgia for its being less than total in admiration; the implicit criticism is tempered with love. Why, it is easy to see. Such characters were, one might say, the eternal

verities of the lost paradise of her childhood, mellowed,
transfigured by the working of memory and by the sense
of what she had lost. These characters for whom she
had such deep affection represented the past. They are,
as it were, fixed, and George Eliot herself was not of
their kind; and in some sense, in the majority of her
novels, certainly in *The Mill on the Floss*, *Felix Holt*
and *Middlemarch* conspicuously (from the very date of
its action, *Daniel Deronda* falls outside these generalisa-
tions), George Eliot opposes to them characters of a
different order of being, characters who can be described
as "free spirits," to use the convenient phrase employed
by C. B. Cox in his book of that name, characters for
whom, as for George Eliot herself, traditional morality,
however admirable, is not enough, who can be satisfied
with no moral code that does not express to the full their
deepest aspirations and their sense of the potentialities
within them.

Their great representatives, of course, are Maggie
Tulliver, in *The Mill on the Floss*,

> . . . a creature full of eager, passionate longings for all
> that was beautiful and glad; thirsty for all knowledge; with
> an ear straining after dreamy music that died away and
> would not come near to her; with a blind, unconscious
> yearning for something that would link together the won-
> derful impressions of this mysterious life, and give her soul
> a sense of home in it . . .

and Dorothea Brooke, in *Middlemarch*, a St. Theresa who
found for herself "no epic life wherein there was a constant unfolding of far-resonant action." Allied to them,
though seen much more from the outside, are such male
characters as Felix Holt, and Lydgate and Ladislaw in
Middlemarch. Simply because they are free spirits, living
by aspiration and—in the realm of conduct—by something like experiment, these free spirits are much more

vulnerable than the heirs of traditional morality such as Mrs. Poyser, Adam Bede and Mr. Garth, who, even when all seems to have gone wrong with their fortunes, can still take comfort in the fact that they have done their duty. Duty is plain to them, as it is not to the free spirits. Because it is not plain, because it may even appear as a choice between duties that seem to conflict, it is the more awful, and the obligation to follow it the more imperative. For the free spirit, indeed, the recognition of where duty lies and what it is is bound up with self-knowledge. "Great feelings," says George Eliot in her final paragraphs on Dorothea Brooke, "will often take the aspect of error, and great faith the aspect of illusion."

It was on the thoroughness and cautiousness of her investigations into the problems of conduct as they face the free spirit, who must be responsible to himself in the absence of traditional and religious sanctions felt as binding, that George Eliot's great moral authority in the nineteenth century rested. The effectiveness of these investigations is due not primarily to the seriousness with which she approached them—that alone would not have availed—but to the skill with which characters and problems are woven into the texture of the society she is describing. As much as Arnold Bennett's, her novels are saturated with the sense of the actual. It has often been pointed out that, far more than most English novels, George Eliot's describe a world at work. Hers is the workaday world in the most literal sense; her characters are shown not "in themselves" alone but always in relation to their professions or trades, which are almost invariably sources of pride and personal fulfilment. For George Eliot and her characters, work, it might almost be said, has a sacramental quality; it is one of the factors that bind them to the community in which they live; it is the duty that is daily nearest to hand.

It is in this context of a working community that

George Eliot dramatises the moral problems of her characters. These, as Joan Bennett has observed, are always concerned with the adjustment of the individual to the community, "with the discovery of a mean point between complete self-repression and unchecked self-indulgence." George Eliot sees society as a closely woven web which, touched at any point, trembles in all its parts. Hence the awful necessity of right action and the avoidance of wrong. Mr. Irwine, in *Adam Bede*, speaks for his creator when he says: "Consequences are unpitying. Our deeds carry their terrible consequences, quite apart from any fluctuations that went before—consequences that are hardly ever confined to ourselves." The words, indeed, define the inner structure of George Eliot's novels as they do her moral view. An act of wrong-doing brings its own retribution down upon the perpetrator, but, beyond this, it affects the whole community; its consequences are incalculable. Arthur Donnithorne's thoughtless seduction of Hetty Sorrel, in *Adam Bede*, not only leads to her murder of her baby and her imprisonment and to Adam's wretchedness, it also brings shame on the Poysers and very nearly their departure from Hayslope. In other words, it disrupts a whole community. Nothing can be the same again.

The consequences of right-doing are necessarily less dramatic but just as important. The final paragraph of *Middlemarch* runs:

> Her finely-touched spirit had still its fine issues, though they were not widely visible. Her full nature, like that river of which Cyrus broke the strength, spent itself in channels which had no great name on the earth. But the effect of her being on those around her was incalculably diffusive: for the growing good of the world is partly dependent on unhistoric acts; and that things are not so ill with you and me as they might have been, is half owing

to the number who lived faithfully a hidden life, and rest in unvisited tombs.

Her view of morality being as stern and intransigent as it is, George Eliot's constant insistence on the necessity and beauty of self-sacrifice cannot surprise. For her, the only guide to right behaviour was the good of others. It is what the characters in her fiction who live according to traditional patterns know, as it were, by instinct; the free spirits have to learn it the hard way.

2

THE THREE STORIES that make up *Scenes of Clerical Life—The Sad Fortunes of the Reverend Amos Barton, Mr. Gilfil's Love Story* and *Janet's Repentance*—need not detain us long. They were prentice-work, and though in the light of the later novels it is easy now to find in them adumbrations of what was to come, it is difficult to imagine how anyone reading them at the time of their publication could have foreseen that a great novelist was in the making. They could scarcely be more clumsily told; the prose is often atrocious; and they indulge, in *Mr. Gilfil's Love Story* and *Janet's Repentance*, much too recklessly in melodrama.

They have their moments, of course. Amos Barton himself is still an affecting character, and one recognises George Eliot's courage in making so drab a figure the hero of a story. Throughout the stories, there is always the honesty we expect of her; and some of the minor characters—the old women, the doctors, the servants—are done well in a way that she was to do better.

This, for instance, from *Amos Barton*, is authentic George Eliot:

> "I never saw the like to parsons," Mr. Hackit said one day in conversation with his brother churchwarden, Mr. Bond; "they're al'ys for meddling with business, an' they know no more about it than my black filly."

"Ah," said Mr. Bond, "they're too high learnt to have much common sense."

Or this, from *Mr. Gilfil's Love Story:*

Mr. Bates was by no means an average person, to be passed without special notice. He was a sturdy Yorkshireman, approaching forty, whose face Nature seemed to have coloured when she was in a hurry, and had no time to attend to *nuances,* for every inch of him visible above his neckcloth was of one impartial redness; so that when he was at some distance your imagination was at liberty to place his lips anywhere between his nose and chin. Seen closer, his lips were discerned to be a peculiar cut, and I fancy this had something to do with the peculiarity of his dialect, which, as we shall see, was individual rather than provincial. Mr. Bates was further distinguished from the common herd by a perpetual blinking of the eyes; and this, together with the red-rose tint of his complexion, and a way he had of hanging his head forward, and rolling it from side to side as he walked, gave him the air of Bacchus in a blue apron, who, in the present reduced circumstances of Olympus, had taken to the management of his own vines. Yet, as gluttons are often thin, so sober men are often rubicond; and Mr. Bates was sober, with that manly, British, churchman-like sobriety which can carry a few glasses of grog without any perceptible clarification of ideas.

And Mrs. Linnett, in *Janet's Repentance,* plainly proclaims herself as of the company of George Eliot's formidable women from Mrs. Poyser to Mrs. Cadwallader when she says:

"I've nothing to say agin' her piety, my dear; but I know very well I shouldn't like her to cook my victual. When a man comes in hungry and tired, piety won't feed him, I reckon. Hard carrots 'ull lie heavy on his stomach, piety or no piety. I called in one day while she was dishing up Mr. Tryan's dinner, an' I could see the potatoes was as watery as watery. It's right enough to be speritial—I'm no enemy to that; but I like my potatoes mealy."

What is remarkable is the tremendous advance on *Scenes of Clerical Life* shown by *Adam Bede*. It has its faults, but in it George Eliot comes into her own. In writing *Adam Bede* she was, of course, going back to family legend:

The germ of "Adam Bede" was an anecdote told me by my Methodist Aunt Samuel (the wife of my father's youngest brother),—an anecdote from her own experience. We were sitting together one afternoon during her visit to me at Griff, probably in 1839 or 1840, when it occurred to her to tell me how she had visited a condemned criminal,—a very ignorant girl who had murdered her child and refused to confess; how she had stayed with her praying through the night, and how the poor creature at last broke into tears and confessed her crime. My aunt afterwards went with her in the cart to the place of execution; and she described to me the great respect with which this ministry of hers was regarded by the official people about the gaol. This story, told by my aunt with great feeling, affected me deeply, and I never lost the impression of that afternoon and our talk together. . . .

The character of Dinah grew out of my recollections of my aunt, but Dinah is not at all like my aunt, who was a very small, black-eyed woman, and (as I was told, for I never heard her preach) very vehement in her style of preaching. She had left off preaching when I knew her, being probably sixty years old, and in delicate health; and she had become, as my father told me, much more gentle and subdued than she had been in the days of her active ministry and bodily strength, when she could not rest without exhorting and remonstrating in season and out of season. . . .

The character of Adam and one or two incidents connected with him were suggested by my father's early life; but Adam is not my father any more than Dinah is my aunt.

There is no reason to doubt this; yet the novel is
saturated with memories of her childhood, or rather of
the life of her family as she learnt of it in her childhood.
These memories, though, come to us mediated, shaped
into art, by the influence of Scott. There are very obvious
similarities of incident with *The Heart of Midlothian*, but
the mark of Scott on the novel goes deeper than this. It
is her most profoundly Tory novel; she is looking back to,
re-creating, a past that, for all the realism with which it
is presented, seems almost ideal, a traditional society in
which everyone knows his place and his duty and is re-
spected accordingly.

One can apply to many of the characters in *Adam
Bede* Bagehot's words on Scott: "A hard life many . . .
seem to lead; but he appreciates, and makes his readers
appreciate, the full value of natural feelings, plain
thoughts, and applied sagacity." There is a sense in
which the Poysers, and Adam Bede himself, come right
out of Scott. One may quote the following passage of
dialogue as wholly characteristic of the Poysers' speech:

"Nay, nay," said Mr. Poyser, "thee mustn't judge Hetty
too hard. Them young gells are like th'unripe grain; they'll
make a good meal by-and-by, but they're squashy as yet.
Thee'll see Hetty'll be all right when she's got a good
husband and children of her own."

"*I* don't want to be hard upo' the gell. She's got cliver
fingers of her own, and can be useful enough when she
likes, and I should miss her wi' the butter, for she's got a
cool hand. An' let be what may I'd strive to do my part by
a niece of yours, and *that* I've done; for I've taught her
everything as belongs to a house, an' I've told her her duty
often enough, though, God knows, I've no breath to spare,
an' that catching pain comes on dreadful by times. Wi'
them three gells in the house I'd need have twice the
strength to keep 'em up to their work. It's like having
roast-meat at three fires; as soon as you've basted one,
another's burnin'."

Their speech is idiosyncratic, and what makes it so are the images they draw on. These images all come from the experience of their daily work; Mr Poyser's are those of a farmer, Mrs. Poyser's those of a housewife. Similarly, Adam Bede thinks and talks continually in terms of his trade as a carpenter. They are individualised beyond this, but all gain strength from the fact that their speech is generic, that of a special category of beings of which each is an individual member. Scott does this with his characters, or at any rate his characters from humble life; one recalls as an example Andrew Fairservice, the gardener in *Rob Roy*. Such characters take on universality by epitomising the general in the specific; they seem to rise out of the enduring norms of life and to represent them. Mrs. Poyser is at once one sharply differentiated farmer's wife and every farmer's wife, almost the Platonic idea of the farmer's wife. If, by comparison, George Eliot fails, as it seems to me she does, with Adam Bede, it is because his role in the action demands that he shall not be a representative carpenter, and his constant analogies from his trade seem at once too apt and too facile; they emerge as copy-book maxims.

From this device comes much of the solidity and depth of George Eliot's re-creation of the pastoral Warwickshire scene, for it emphasises the enduring aspects of life. But she has another way also of emphasising these enduring aspects, one that is brilliantly analysed by Dorothy Van Ghent in her *The English Novel: Form and Function*. This is through what Professor Van Ghent calls "the massively slow movement" of the novel, which moves to the time of Mrs. Poyser's eight-day clock, in other words to the rhythms of the farming life. George Eliot can afford to be leisurely because the life of the society she is describing is leisurely. It is this society—symbolised by the Hall Farm and all its activities: its interior "where the only chance of collecting a few grains of dust would be to climb on the salt-coffer, and put your finger on the high mantel-shelf"; its oak table polished with genuine "elbow polish"; its dairy "a scene to sicken for with a sort of calenture in hot and dusty streets—such coolness, such purity, such fresh fragrance of new-pressed cheese, of firm butter, of wooden vessels perpetually bathed in pure water"; its inhabitants the Poysers, and not least among them old Martin Poyser, a "hale but shrunken and bleached image of his portly, black-haired son . . . watching what went forward with the quiet *outward* glance of healthy old age"—it is this society that Arthur Donnithorne's thoughtless act all but disrupts.

The seduction of Hetty is the inner action of the novel. In many respects, it is beautifully done. The rendering of Arthur, the young squire, is especially good, as one realises right away when one compares him, as a representation of the average good-hearted tolerably well-meaning sensual young man, with Thackeray's Pendennis, who is scarcely allowed a sensual life at all, or when one compares the seduction itself with Steerforth's seduction of Little Em'ly in *David Copperfield*. It is shown as a touching, idyllic love-affair, in which Arthur and Hetty are alike caught. The rendering of Hetty is almost a great triumph; certainly the account of her despairing return from Windsor when pregnant is one of the finest things in English fiction; and her pretty sensuality is wonderfully well suggested; it communicates itself to all the other characters in the novel and to the reader as well.

But one has to say "almost a great triumph" because her creator cannot herself accept her creation simply as the child-like suffering girl she is. She is infinitely pathetic, and the full, desperate situation in which she finds herself could surely have been left to make its point. It has to be used, however, again and again to make George Eliot's points, and it is difficult not to feel that George Eliot is guilty of vindictiveness towards her character. It is almost as though Hetty's very prettiness is scored up as a bad mark against her. It is partly, of course, a want of tact on George Eliot's part, the result of a defective technique; she falls into no such gross errors in her later work. Even so, when one comes across this, for instance:

> Yes, the actions of a little trivial soul like Hetty's, struggling amidst the serious, sad destinies of a human being, *are* strange. So are the motions of a little vessel without ballast tossed about on a stormy sea, how pretty it looked with its parti-coloured sail in the sunlight, moored in the quiet bay!
> "Let that man bear the loss who loosed it from its moorings."
> But that will not save the vessel—the pretty thing that might have been a lasting you.

one is outraged not primarily by the authorial intrusion but by something much more serious—the author's plain lack of charity. D. H. Lawrence's belief that we should trust the tale rather than its teller is not universally valid, but it certainly applies to this side of *Adam Bede*.

George Eliot's method with Adam Bede and Dinah Morris is almost the obverse of her treatment of Hetty, and Adam and Dinah are the less convincing because of that. They are conceived in and presented with almost total admiration. No one has ever been quite persuaded of their marriage; one can only think that since each is much too good for any other character in the novel, there is nothing for it except they should be paired off at the end. It is, of course, in their goodness that the difficulty lies. They are too good to be true; with them, George Eliot over-reaches herself. It is not that one does not believe in Adam; it is rather that one's reaction towards him is the opposite of what George Eliot intends it to be. In his unutterable rectitude, to which even his own

awareness of his faults somehow contributes, he is a
sententious, loquacious prig of a kind that in life one
would walk a long way to avoid. And he is all the time
too palpably his author's spokesman:

"Nay, Seth, lad; I'm not for laughing at no man's religion.
Let 'em follow their consciences, that's all. Only I think it
'ud be better if their consciences 'ud let 'em stay quiet i'
the church—there's a deal to be learnt there. And there's
such a thing as being over-speritual; we must have something beside Gospel i' this world. Look at the canals, an'
the aqueducs, an' the coal-pit engines, and Arkwright's
mills there at Cromford; a man must learn summat beside
Gospel to make them things, I reckon. But t'hear some o'
them preachers, you'd think as a man must be doing nothing
all's life but shutting's eyes and looking what's a-going-on
inside him. I know a man must have the love o' God in his
sould, and the Bible's God's word. But what does the Bible
say? Why, it says as God put His sperrit into the workman
as built the tabernacle, to make him do all the carved work
and things as wanted a nice hand. And this is my way o'
looking at it; there's the sperrit o' God in all things and all
times—weekday as well as Sunday—and i' the great works
and inventions, and i' the figuring and the mechanics. And
God helps us with our head-pieces and with our hands as
well as with our souls; and if a man does bits o' jobs out
o' working hours—builds a oven for's wife to save her from
going to the bakehouse, or scrats at his bit o' garden and
makes two potatoes grow instead o' one, he's doing more
good, and he's just as near to God, as if he was running
after some preacher and a-praying and a-groaning."

The passage is a notable piece of ventriloquialism on
George Eliot's part, and what is said is excellent doctrine; but it is primarily hers, not Adam's, and is felt
to be such.

The problem with Dinah is rather different. *Adam
Bede* is a superb study of the impact of Methodism on
lower-class English life, and it is in the figure of Dinah

that this aspect of the novel is contained. She is shown always in the very actions of goodness, but conscious goodness is the most difficult quality a novelist can portray, no doubt because, though it undoubtedly exists, we know, and all the saints have been agonisingly aware, that it cannot exist unalloyed. Dinah is too much a model of religious and moral excellence to be convincing as a human being. As an artist, George Eliot seems to have felt this herself, for time and again she applies a corrective to Dinah by juxtaposing her with Mrs. Poyser. She is given a foil as Adam is not:

". . . But where's the use o' talking, if ye wonna be persuaded, and settle down like any other woman in her senses, instead o' wearing yourself out with walking and preaching, and giving away every penny you get, so as you've saved nothing against sickness; and all the things you've got i' the world, I verily believe, 'ud go into a bundle no bigger nor a double cheese. And all because you've got notions i' your head about religion more nor what's i' the Catechism and the Prayer-book."

"But not more than what's in the Bible, aunt," said Dinah.

"Yes, and the Bible too, for that matter," Mrs. Poyser rejoined, rather sharply; "else why shouldn't them as know best what's in the Bible—the parsons and people as have got nothing to do but learn it—do the same as you do? But, for the matter o' that, if everybody was to do like you the world must come to a standstill; for if everybody tried to do without house and home, and with poor eating and drinking, and was allays talking as we must despise the things o' the world, as you say, I should like to know where the pick o' the stock, and the corn, and the best new-milk cheese 'ud have to go. Everybody 'ud be wanting bread made o' tail ends, and everybody 'ud be running after everybody else to preach to 'em, instead o' bringing up their families, and laying by against a bad harvest. It stands to sense as that can't be the right religion. . . ."

Mrs. Poyser here speaks as the representative of common sense, inadequate though it may be, of traditional values, as indeed it is her function to do throughout the novel.

The resolution of the novel—Hetty's reprieve at the last minute; Arthur's return, purged by suffering; the marriage of Adam and Dinah—has never satisfied anybody, and Henry James was probably right when he doubted whether George Eliot herself "had a clear vision" of them, adding: "They are a very good example of the view in which a story must have marriages and rescues in the nick of time, as a matter of course." Yet, when all has been said against it, *Adam Bede* remains a great achievement, memorable for the tragedy of Hetty and, beyond this, for the ambience in which it is enacted, the pastoral circle of life in Hayslop that encloses the moral action, the scenes at the Hall Farm and the rectory, Arthur Donnithorne's twenty-first-birthday celebrations, the harvest supper and the rest. In these, the endless rhythm of one kind of "monotonous homely existence" is caught for ever, in detail whose massiveness cannot disguise its sharpness. As Dorothy Van Ghent well says, the protagonist in this novel is the community itself. It is this that makes *Adam Bede* the finest pastoral novel in English, *Far from the Madding Crowd* not excepted.

3

G EORGE ELIOT's fiction falls into two parts. *Scenes of Clerical Life, Adam Bede, The Mill on the Floss* and *Silas Marner* were all published between 1858 and 1861. The fiction of her second period was in some respects more ambitious; it opened with *Romola* in 1864, which was followed by *Felix Holt, the Radical* in 1866, *Middlemarch* in 1871–1872 and *Daniel Deronda* in 1876.

What is remarkable is the speed with which, having discovered she could write fiction, she produced her early novels; it shows how accessible was the vein of imagination on which she drew. It was composed very largely of her own memories of childhood and of what may be called family legend. She was to write in *Daniel Deronda:*

> A human life, I think, should be well rooted in some spot of a native land, where it may get the love of tender kinship for the face of the earth, for the labours men go forth to, for the sounds and accents that haunt it, for whatever will give that early home a familiar unmistakable difference amidst the future widening of knowledge: a spot where the definiteness of early memories may be inwrought with affection, and kindly acquaintance with all neighbours, even to the dogs and donkeys, may spread not by sentimental effort and reflection, but as a sweet habit of the blood. At five years old, mortals are not prepared to be citizens of the world, to be stimulated by abstract nouns, to soar above preference into impartiality; and that prejudice in

favour of milk with which we blindly begin, is a type of the way body and soul must get nourished at least for a time. The best introduction to astronomy is to think of the nightly heavens as a little lot of stars belonging to one's own homestead.

This deeply rooted, severely localised life of childhood was something Gwendolen Harleth lacked and that her creator had had to the full. To it she returned in *The Mill on the Floss*. It is obviously not an autobiographical novel in anything like a complete sense, as, for example, *David Copperfield* and, especially, *Sons and Lovers* and *A Portrait of the Artist as a Young Man* are. Geographically, St. Ogg's is a considerable distance from Arbury, and a river of the amplitude and power of the Floss is conspicuously missing from the north Warwickshire landscape. And certainly Mr. Tulliver is not a portrait of Robert Evans. On the other hand, however different from George Eliot's are the circumstances in which she is placed, Maggie Tulliver is in all essentials a close projection of the author, so that it is true to see the novel as George Eliot's *À la recherche du temps perdu*. We know that Proust was much influenced by George Eliot, and we see, as Joan Bennett has said, the adult figures in the novel both as the child sees them and as the mature artist comprehends them, as in *Du côté de chez Swann*.

This identification of author with heroine, though it does not at all preclude criticism of the heroine by the author, gives *The Mill on the Floss* a sensuous warmth and a personal urgency beyond anything found in *Adam Bede*. As a rendering of the growth of a girl from early childhood to young womanhood, a girl marked by intellectual distinction, a generously ardent nature and a strong capacity for feeling, Maggie has never been surpassed. She speaks for herself at all times when she says to Philip Wakem: "I was never satisfied with a *little* of anything. That is why it is better for me to do without earthly happiness altogether. . . . I never felt that I had enough music—I wanted more instruments playing together—I wanted voices to be fuller and deeper." She has, indeed, an excess of sensibility, almost an excess of expectation; she is too ardent, swerving passionately from the extreme of desire to the extreme of self-abnegation. As Philip tells her, speaking for George Eliot but speaking all the same in character:

". . . you are shutting yourself up in a narrow self-delusive fanaticism, which is only a way of escaping pain by starving into dullness all the highest powers of your nature. Joy and peace are not resignation; resignation is the willing endurance of a pain that is not allayed—that you don't expect to be allayed. Stupefaction is not resignation; and it is stupefaction to remain in ignorance—to shut up all the avenues by which the life of your fellow-men might become known to you. I am not resigned: I am not sure that life is long enough to learn that lesson. *You* are not resigned: you are only trying to stupefy yourself."

And again:

"Maggie," he said, in a tone of remonstrance, "don't persist in this wilful, senseless privation. It makes me wretched to see you benumbing and cramping your nature in this way. You were so full of life when you were a child: I thought you would be a brilliant woman—all wit and bright imagination. And it flashes out in your face still, until you draw that veil of dull quiescence over it. . . . It is mere cowardice to seek safety in negations. No character becomes strong in that way. You will be thrown into the world some day, and then every rational satisfaction of your nature that you deny now, will assault you like a savage appetite."

That last sentence is prophetic of the tragic plight into which Maggie is to fall; but what is interesting at this moment is the criticism, which one is bound to see as

George Eliot's criticism of herself, made by the only character in the book with sensitivity and intellect enough to understand Maggie. Yet the criticism of her tendency towards excess is implicit, it seems to me, in descriptions of her behaviour from early childhood. She is heart-broken when, as a child, her brother Tom moves away from her, as in the nature of things he must. This tragedy of childhood is beautifully done; and yet one can't help feeling that Maggie's demands on Tom were in their nature excessive. When Philip sees her at Mr. Stelling's, her dark eyes reminded him of the "stories about princesses being turned into animals." He wonders why the thought should come to him, and George Eliot herself gives the answer: "I think it was that her eyes were full of unsatisfied intelligence, and unsatisfied, beseeching affection." What seems clear is that Maggie's affection is of a kind so beseeching in its nature as to be incapable of satisfaction, at any rate in the brutal society in which she must live.

For the society George Eliot describes, that of St. Oggs, and the Tullivers and the ramifications of the Dodson family, *is* brutal, and Maggie's tragedy is that of the free spirit caught in a blankly materialistic world. It is a world ruled over entirely by the sense of property, by self-regard and by pride in family; but family itself is valued more as a vehicle for the preservation and transmission of property than for any other reason. George Eliot's descriptions of the Dodson ladies and their husbands are probably the high points in her comedy; they are done with a warm appreciation of idiosyncrasy, in which there is certainly affection.

In a chapter of authorial intervention, George Eliot defends the Tullivers and Dodsons, not, it seems to me, too convincingly. "It is a sordid life, you say, this of the Tullivers and Dodsons . . ." George Eliot does her best for them, but, the fact remains, it *is* a sordid life, whether

one compares it with that of the Poysers in *Adam Bede*
or even of the Vincys in *Middlemarch*. It is rendered
tolerable, one sometimes feels, only because we see it
refracted through the author's humour:

Few wives were more submissive than Mrs. Tulliver on all
points unconnected with her family relations; but she had
been a Miss Dodson, and the Dodsons were a very respectable family indeed—as much looked up to as any in their
parish, or the next to it. The Miss Dodsons had always
been brought up to hold their heads very high, and no one
was surprised the two eldest had married so well—not at
an early age, for that was not the practise of the Dodsons.
There were particular ways of doing everything in that
family: particular ways of bleaching the linen, of making
the cowslip wine, curing hams, and keeping the bottled
gooseberries; so that no daughter of that house could be
indifferent to the privilege of having been born a Dodson,
rather than a Gibson or a Watson. Funerals were always
conducted with peculiar propriety in the Dodson family:
the hat-bands were never of a blue shade, the gloves never
split at the thumb, everybody was a mourner who ought
to be, and there were always scarfs for the bearers. When
one of the family was in trouble or sickness, all the rest
went to visit the unfortunate member, usually at the
same time, and did not shrink from uttering the most
disagreeable truths that correct family feeling dictated:
if the illness or trouble was the sufferer's own fault, it
was not in the practice of the Dodson family to shrink from
saying so. In short, there was in this family a peculiar
tradition as to what was the right thing in household
management and social demeanour, and the only bitter
circumstances attending this superiority was a painful inability to approve the condiments or the conduct of families ungoverned by the Dodson tradition. A female Dodson,
when in "strange houses" always ate dry bread with her
tea, and declined any sort of preserves, having no confidence in the butter, and thinking that the preserves had
probably begun to ferment from want of due sugar and

boiling. There were some Dodsons less like the family than others—that was admitted; but in so far as they were "kin," they were of necessity better than those who were of "no kin." And it is remarkable that while no individual Dodson was satisfied with any other individual Dodson, each was satisfied, not only with him or her self, but with the Dodsons collectively. The feeblest member of a family —the one who has the least character—is often the merest epitome of the family habits and traditions; and Mrs. Tulliver was a thorough Dodson, though a mild one, as small-beer, so long as it is anything, is only describable as very weak ale: and though she had groaned a little in her youth under the yoke of her elder sisters, and still shed occasional tears at their sisterly reproaches, it was not in Mrs. Tulliver to be an innovator on the family ideas. She was thankful to be a Dodson, and to have one child who took after her own family at least in his features and complexion, in liking salt and in eating beans, which a Tulliver never did.

Mrs. Tulliver is the mildest of the Dodson sisters, but it is characteristic of her that, faced with her husband's bankruptcy and the consequences of his stroke, her deepest response should be:

"Oh dear, oh dear," said Mrs. Tulliver, "to think o' my chany being sold i' that way—and I bought it when I was married, just as you did yours, Jane and Sophy: and I know you didn't like mine, because o' the sprig, but I was fond of it; and there's never been a bit broke, for I've washed it myself—and there's the tulip on the cups, and the roses, as anybody might go and look at 'em for pleasure. You wouldn't like *your* chany to go an old song and be broke to pieces, though yours 'as got no colour in it, Jane—it's all white and fluted, and didn't cost so much as mine. And there's the casters—sister Deane, I can't think but you'd like to have the casters, for I've heard you say they're pretty."

Equally characteristic are the responses of her sisters to this:

"Why, I've no objection to buy some of the best things,"
said Mrs. Deane, rather loftily; "we can do with extra
things in our house."

"Best things!" exclaimed Mrs. Clegg with severity, which
had gathered intensity from her long silence. "It drives me
past patience to hear you all talking o' best things, and
buying in this, that, and the other, such as silver and
chany. You must bring your mind to your circumstances,
Bessy, and not be thinking o' silver and chany; but whether
you shall get so much as a flock-bed to lie on, and a blanket
to cover you, and a stool to sit on. You must remember, if
you get 'em, it'll be because your friends have bought 'em
for you, for you're dependent upon *them* for everything; for
your husband lies there helpless, and hasn't got a penny i'
the world to call his own. And it's for your own good I say
this, for it's right you should feel what your state is, and
what disgrace your husband's brought on your own family,
as you've got to look to for everything—and be humble in
your mind."

George Eliot has never been praised enough for her
grasp of the property basis of the society she describes
in *The Mill on the Floss*. It is a society in which generosity itself is a vice. Mr. Tulliver is not much better
than the Dodsons; his downfall is the result of his bull-
like obstinacy, but it is made the more certain by his act
of generosity to his sister Grittie; and significantly, George
Eliot shows how this act is related to his love for his
daughter: "It had come across his mind that if he were
hard upon his sister, it might somehow tend to make
Tom hard upon Maggie at some distant day, when her
father was no longer there to take her part; . . . this was
his confused way of explaining to himself that his love
and anxiety for 'the little wench' had given him a new
sensibility towards his sister." Ironically, it is a factor in
his undoing. If one wants a personification of spontaneous goodness, uncalculating sympathy in the novel,
one has to turn to a character of whom not much is

thought by anyone, who scarcely exists in Dodson and
Tulliver terms—the pedlar Bob Jakin. The best that can
be said of Tom Tulliver is that he will grow up to be a
man his Dodson aunts and their husbands can be proud
of.

Such, then, are the society and the circumstances in
which Maggie must have her being. George Eliot's
description of them is masterly, both in detail and in the
large set-pieces of family confrontations and conferences.
But we see them, of course, always in relation to Maggie:

> Maggie in her brown frock, with her eyes reddened and
> her heavy hair pushed back, looked from the bed where her
> father lay, to the dull walls of this sad chamber which
> was the centre of her world, was a creature full of eager,
> passionate longings for all that was beautiful and glad;
> thirsty for all knowledge; with an ear straining after dreamy
> music that died away and would not come near to her; with
> a blind, unconscious yearning for something that would link
> together the wonderful impressions of this mysterious life,
> and give her soul a sense of home in it.
>
> No wonder, when there is this contrast between the out-
> ward and the inward, that painful collisions come of it.

Maggie, since she is the girl she is, living in the society
in which she does, is a figure doomed to tragedy.

The resolution of the tragedy, however, is another
matter and, since the novel's initial publication, has
satisfied no one. The resolution falls into two parts: there
is first Maggie's running away with Stephen Guest, only
to renounce him, and then the final flood scenes in which
she rescues Tom from the mill, only for them to be
carried to their death together.

It is obviously right that Maggie should be swept away
by the passionate intensity of her nature that she has for
too long unnaturally repressed. Indeed, it is prepared
for deliberately by Philip Wakem's prediction: "You will
be thrown into the world some day, and then every

rational satisfaction of your nature that you deny now, will assault you like a savage appetite." It is also obviously right that she should misconstrue her sympathy and fellow-feeling for Philip as love for him and then, having privately committed herself to him, fall in love with another man. The other man, Stephen Guest, who is privately committed to her cousin Lucy, as Maggie knows, has been constantly assailed by critics. "A mere hairdresser's block," Leslie Stephen called him; and he aroused Swinburne's notable powers of denunciation—a "cur" for whom horsewhipping was too good. Admittedly, the first impression he makes on us is of being rather a bounder, certainly of being a too self-satisfied, consciously "superior," facile young man.

George Eliot, in fact, presents him at the beginning in ironical terms. But Stephen's and Swinburne's reactions to him seem excessive; the result, one can't help thinking, of a misreading of Maggie, a failure to heed the criticisms of her made time and again in the course of the novel. In any case, there is nothing in the nature of things to prevent even so fine a spirit as Maggie falling in love with a spiritually coarse young man. These things happen; and, in fact, Joan Bennett is almost certainly right when she concludes that George Eliot intended Stephen to be changed, "improved," by his love for Maggie, to discover, as a result, a new sincerity, new depths, within himself. If this is not the impression we take away from reading the novel, the fault must lie where I believe it does—in George Eliot's inadequate technique.

Stephen Guest appears very late in the novel, a newcomer almost out of the blue and also of an order different from that to which all the other characters belong. He is of another and higher social class than that which we have met before. We are unprepared for him, and George Eliot, who needed space in which to build up her major characters, has not allowed space or time in which to establish him. We have not been given the necessary opportunity to get used to him, to know him; he is a rush-job.

Maggie's renunciation of him is another matter, and one can only feel that she has become the victim of what may be called an either-or morality that her author has imposed upon her and that denies the facts of experience as known at any time in history. George Eliot has almost gone out of her way to ensure that there has been no public engagement between either Maggie and Philip or Lucy and Stephen. Neither couple, in the eyes of the world, is irrevocably committed. The moral appeal, therefore, is to something beyond the conventional or the legalistic. When she renounces Stephen, Maggie is not sacrificing herself to save a marriage, and her self-sacrifice will not decrease the total stock of the unhappiness of the four people involved. Philip and Lucy will be unhappy in any event; her behaviour merely guarantees that she and Stephen must also be unhappy.

The situation in which Maggie and Stephen find themselves in is not, after all, an uncommon one and can never have been. It is one that always involves someone's unhappiness; it is not improved by behaving in such a way as to multiply the unhappiness by two. For the situation is one in which an either-or morality cannot apply, if it can in any situation. George Eliot really sacrifices Maggie to her own doctrine of the virtue of self-sacrifice for its own sake. And the mind revolts against the cruelty implicit in it—and the falsity as well.

On the face of it, the *dénouement*, Maggie's rescue of her estranged brother in the flood and their reunion in death, is well prepared. The Floss flows through the novel from beginning to end. It is there in the title of the novel; it is in the very first sentence as in the last paragraphs. It may be said to encompass and enclose the action. It is closely bound up with the Tullivers' downfall:

"There's a story," Mr. Tulliver says, "as when the mill changes hands, the river's angry—I've heard my father say it many a time." As for Mrs. Tulliver: " 'They're such children for the water, mine are,' she said aloud, without reflecting that there was no one to hear her; 'They'll be brought in dead and drowned some day. I wish that river was far enough.' " The flood itself may be seen as a retrospective symbol of the flood of passion that has swept Maggie far from the narrow confines of her normal life and conduct.

The flood was obviously in George Eliot's mind from the beginning; we find her writing in her journal, on January 12, 1859: "We went into town today and looked in the 'Annual Register' for cases of *inundation*." Yet, as the *dénouement* of the novel, it cannot help but strike one as a quite artificial resolution. It is as though Maggie, having made her final act of self-sacrifice, has nothing left for her but death. Even so, the manner of the death is fundamentally unsatisfying, as the very prose in which it is reported, in the last paragraph of the novel, indicates:

> The boat reappeared—but brother and sister had gone down in an embrace never to be parted: lived through again in one supreme moment, the days when they had clasped their little hands in love, and roamed the daisied fields together.

That is sentimentality of the most flagrant kind, as indeed is the notion that Maggie and Tom can ever be united in any real sense again. The return to childhood and to the security of childhood with Tom simply does not work. Yet one sees George Eliot's difficulty. The *dénouement* was forced upon her, one feels, by the falsity of Maggie's renunciation of Stephen. But there is something else implicit in it beyond this. If one looks at later autobiographical novels, Bennett's *Clayhanger*, Joyce's *A Portrait of the Artist as a Young Man*, Lawrence's *Sons and Lovers*, one realizes that in an old-fashioned sense they do not end. There can be no resolution because no action has been completed; a phase of development has been described—no more. Instead of an ending as such, we have a change in direction, as it were, a departure to a new scene: Dedalus's "O life! I go to encounter for the millionth time the reality of experience and to forge in the smithy of my soul the uncreated conscience of my race"; Paul Morel's "But no, he would not give in. Turning sharply, he walked towards the city's gold phosphorescence. His fists were shut, his mouth set. He would not take that direction, to the darkness, to follow her. He walked towards the faintly humming, glowing town, quickly."

But nothing like this was possible for George Eliot at this stage in her career and at the date at which she was writing. She could only fall back, however brilliantly she handled it, on one of the cliché endings of Victorian fiction.

4

G EORGE ELIOT wrote *Silas Marner, the Weaver of Raveloe* between November, 1860, and March, 1861; it
"has thrust itself between me and the other book I was
meditating" (*Romola*).

Though slight, *Silas Marner* is as perfect a work of
prose fiction as any in the language, a small miracle. It
is a book that in her own time only George Eliot could
have brought off and that, since her day, no English or
American novelist would have dared attempt. She herself
described its especial quality, writing to her publisher
Blackwood, as well as anyone has ever done:

> I don't wonder at your finding my story, as far as you
> have read it, rather sombre: indeed, I should not have
> believed that anyone would have been interested in it but
> myself (since Wordsworth is dead) if Mr. Lewes had not
> been strongly arrested by it. But I hope you will not find
> it at all a sad story, as a whole, since it sets—or is intended
> to set—in a strong light the remedial influences of pure,
> natural human relations. The Nemesis is a very mild one.
> I have felt all through as if the story would have lent
> itself best to metrical rather than to prose fiction, especially
> in all that relates to the psychology of Silas; except that,
> under that treatment, there could not be an equal play of
> humour. It came to me first quite suddenly, as a sort of
> legendary tale, suggested by my recollection of having
> once, in early childhood, seen a linen weaver with a bag

on his back; but as my mind dwelt on the subject, I became inclined to a more realistic treatment.

Fortunately, George Eliot did not attempt to write *Silas Marner* in verse, for, though she wrote two volumes of verse, she was no poet; all that are still remembered of her verses are the lines, in Wordsworthian blank verse, from *O May I Join the Choir Invisible,* which were much anthologised during the nineteenth century, and the sonnet sequence *Brother and Sister*—and that for its possible autobiographical connotations rather than for its quality as poetry.

All the same, though she could not achieve poetry when writing verse, she did so triumphantly in the prose of *Silas Marner;* it lives in the memory as a poem—and as a poem that one feels George Eliot was right in suggesting that, herself apart, only Wordsworth could have written. "The remedial influences of pure, natural human relations"—the concept is Wordsworthian, and the work as a whole is George Eliot's profoundest and most beautiful expression of what she herself called, as a very young woman, "the truth of feeling."

And though, as she began to write it, she gave her story "a more realistic treatment," it remains a "sort of legendary tale." Indeed, the note of legend, of events occurring in a past already remote, is struck in the very first sentence of the story: "In the days when the spinning wheels hummed busily in the farmhouses—and even great ladies, clothed in silk and thread lace, had their toy spinning wheels of polished oak—there might be seen, in districts far away among the lanes, or deep in the bosom of the hills, certain pallid undersized men who, by the side of the brawny countryfolk, looked like the remnants of a disinherited race."

So strong is the evocation of remoteness, even of something like the timelessness of fairyland, that it comes as a shock to realise that the action narrated took place in a generation that ended within only a few years of the author's own birth. The events described are not dated with any exactitude, but they cover about thirty years on the very eve of the Victorian age. Yet, even so, they belong, and did when George Eliot was writing the story, to a period as irretrievably in the past as Shakespeare's England.

What had happened, of course, was the coming of the railway. As a modern historian has said: "When the Duke of Wellington attended the opening of the new Manchester and Liverpool Railway in September, 1830, he witnessed an event as important in its own way as the Battle of Waterloo, which he had won fifteen years before. It symbolised the conquest of space and parochialism." It took the Industrial Revolution everywhere; it meant the disappearance of villages like Raveloe, the tempo of whose existence had remained almost unchanged for centuries. The "districts far away among the lanes, or deep in the bosom of the hills" were far away no longer but wide open to the influence of the great world. By 1860, the story of Silas Marner would have been impossible, for the very naïvety of the characters —villagers, gentry and Marner alike—and their rusticity would have been impossible.

It is this naïvety, this rusticity, along with the remoteness of the time and the place, that gives *Silas Marner* its charm. The remoteness also made it possible for George Eliot to tell a story of the utmost simplicity.

Silas Marner is essentially a myth of spiritual rebirth. Marner, the Methodist weaver, pallid, undersized, a child of the dark, satanic mills of the Industrial Revolution, loses his faith when he is accused of and found guilty by his fellow-Methodists of a particularly mean theft. Almost crazed by despair, he leaves the town and wanders through the country, finally settling on the edge of a village worlds away from the Industrial Revolution, unaffected by it, and therefore worlds away from life as he has known it.

The difference between Raveloe and the town of his origins is so great, indeed, as to make communication between him and his new neighbours all but impossible. To the villagers, it did not seem certain that "this trade of weaving, indispensible though it was, could be carried on entirely without the help of the Evil One." The sound of Silas's loom, not far from the edge of a deserted stone pit, is "questionable" and has "a half-fearful fascination for the Raveloe boys." His gaze was "enough to make them take to their legs in terror. For how was it possible for them to believe that those large brown protuberant eyes in Silas Marner's pale face really saw nothing very distinctly that was not close to them, and not rather that their dreadful stare could dart cramp, or rickets, or a wry mouth at any boy who happened to be in the rear?"

Unwittingly, Marner is caught in the "strange lingering echoes of the old demon worship" of the villagers. His very strangeness makes him an object of superstitious fear, and for want of an alternative, he becomes a miser. Then, mysteriously, he is robbed of his hoard of gold, which in due time is, as it were, magically replaced by a golden-haired baby girl whom he finds on his hearth. To the child, Eppie, he devotes himself as ardently as before to his gold. It is as though the precious metal has been transmuted into human affection, and one result of this is that he is accepted without question by the community of Raveloe.

This is plainly of the stuff of myth, and when the followers of Jung discover *Silas Marner*, they will have a field day. We are in the presence of the magical; but the magical is mediated for us by the superstition of the villagers, for whom magic is still a reality.

Silas Marner's affinities to the fairy-tale are obvious enough, but it is a fairy-tale saturated in the sense of the actual. It is this that gives it its enduring power to compel belief. There is Raveloe itself:

> . . . orchards looking lazy with neglected plenty; the large church in the wide churchyard, which men gazed at lounging at their own doors in service time; the purple-faced farmers jogging along the lanes or turning in at the Rainbow; homesteads, where men supped heavily and slept in the light of the evening hearth, and where women seemed to be laying up a stock of linen for the life to come.

It exists in sharp contrast to the industrial world of Lantern Yard, from which Marner has emerged. "In the early ages of the world, we know, it was believed that each territory was inhabited and ruled by its own divinities, so that a man could cross the bordering heights and be out of reach of his native gods, whose presence was confined to the streams and the groves and the hills among which he had lived from his birth." The author's comment indicates the primitive quality of life in Raveloe, which is self-contained and self-sufficient. To it, in a curious way throughout the book, Marner is marginal. He is the character who must convince us—the villagers are there in all their actuality, palpable creatures of flesh and blood and rich idiosyncratic speech; but George Eliot handles Marner with consummate skill.

His presence broods over the novel, but his appearances in it are anything but continuous; he is, in fact, offstage more often than he is on. We see him with a double vision: as the lost, crazed, pathetic being George Eliot presents and also as the mysterious stranger he is to the villagers. But it is the villagers who are the norm of life in the novel; they are created with great warmth and affection, and when they take Marner into their fellowship, we accept him also.

In a sense—beautifully differentiated from one another though the villagers are—Raveloe is itself a character in the novel, a corporate entity, as it were, with its own personality. Part of it is indicated in the following:

> The inhabitants of Raveloe were not severely regular in their church-going, and perhaps there was hardly a person in the parish who would not have held that to go to church every Sunday in the calendar would have shown a greedy desire to stand well with Heaven, and get an undue advantage over their neighbours—a wish to be better than the "common run" that would have implied a reflection on those who had had godfathers and godmothers as well as themselves, and had an equal right to the burying-service.

The entity that is Raveloe has its wisdom; but it is a folk-wisdom, a lore inherited from time immemorial, the expression of a way of life that has scarcely changed over generations. The villagers speak, as it were, in chorus, antiphonally, as in the wonderful scenes in the Rainbow Inn and the interchanges between Mr. Macey and Mr. Winthrop, Mr. Tookey, the butcher and the farrier and Mr. Snell, the landlord, scenes that take us back to Shakespeare's comic rustics and anticipate Hardy's:

> "Come, come," said the landlord; "let the cow alone. The truth lies atween you: you're both right and both wrong, as I always say. And as for the cow's being Mr. Lammeter's, I say nothing to that; but this I say, as the Rainbow's the Rainbow. And for the matter o' that, if the talk is to be o' the Lammeters, you know the most o' that head, eh, Mr. Macey? You remember when first Mr. Lammeter's father come into these parts, and took the Warrens?"
>
> Mr. Macey, tailor and parish clerk, the latter of which functions rheumatism had of late obliged him to share with a small-featured young man who sat opposite him, held his white head on one side, and twirled his thumbs with an

air of complacency, slightly seasoned with criticism. He smiled pityingly, in answer to the landlord's appeal, and said:

"Ay, ay; I know; but I let other folks talk. I've laid by now, and gev up to the young uns. Ask them as have been to school at Tarley: they've learnt pernouncing; that's come up since my day."

"If you're pointing at me, Mr. Macey," said the deputy clerk, with an air of anxious propriety, "I'm nowise a man to speak out of my place. As the psalm says:

'I know what's right, nor only so,
But also practice what I know.'"

"Well, then, I wish you'd keep hold o' the tune when it's set for you; if you're for prac*tis*ing, I wish you'd prac*tise* that," said a large, jocose-looking man, an excellent wheelwright in his weekday capacity, but on Sundays leader of the choir. He winked, as he spoke, at two of the company, who were known officially as the "bassoon" and the "key bugle," in the confidence that he was expressing the sense of the musical profession in Raveloe.

Mr. Tookey, the deputy clerk, who shared the unpopularity common to deputies, turned very red, but replied, with careful moderation, "Mr. Winthrop, if you'll bring me any proof as I'm in the wrong, I'm not the man to say I won't alter. But there's people set up their own ears for a standard, and expect the whole choir to follow 'em. There may be two opinions, I hope."

"Ay, ay," said Mr. Macey, who felt very well satisfied with this attack on youthful presumption; "you're right there, Tookey: there's always two 'pinions; there's the 'pinion a man has of hissen, and there's the 'pinion other folks have on him. There'd be two 'pinions about a cracked bell, if the bell could hear itself."

"Well, Mr. Macey," said poor Tookey, serious amidst the general laughter, "I undertook to partially fill up the office of parish clerk by Mr. Crackenthorp's desire, whenever your infirmities should make you unfitting; and it's one of the rights thereof to sing in the choir—else why have you done same yourself?"

"Ah, but the old gentleman and you are two folks," said Ben Winthrop. "The old gentleman's got a gift. Why, the Squire used to invite him to take a glass, only to hear him sing 'The Red Rover'; didn't he, Mr. Macey? It's a natural gift. There's my little lad Aaron, he's got a gift—he can sing a tune straight off—like a throstle. But as for you, Master Tookey, you'd better stick to your 'Amens': your voice is well enough when you keep it up your nose. It's your inside as isn't right made for music: it's no better nor a hollow stalk."

In Raveloe the right opinion is the opinion other people have, the community opinion. But there is something else to be noted about the community that is Raveloe: it speaks with one voice, one accent. This comes out in the speech of the chorus of ladies preparing for the dance at the Red House, in Chapter 11, which in its way balances the male chorus at the Rainbow:

"Don't talk *so*," said Nancy, blushing. "You know I don't mean ever to be married."

"Oh, you never mean a fiddlestick's end!" said Priscilla, as she arranged her discarded dress, and closed her band-box. "Who shall *I* have to work for when Father's gone, if you are to go and take notions in your head and be an old maid, because some folks are no better than they should be? I haven't a bit o' patience with you—sitting on an addled egg forever, as if there was never a fresh un in the world. One old maid's enough out o' two sisters; and I shall do credit to a single life, for God A'mighty meant me for it. Come, we can go down now. I'm as ready as a mawkin *can* be—there's nothing a-wanting to frighten the crows, now I've got my ear droppers in."

The point is that the Misses Lammeter are, in Raveloe terms, fine ladies; but they speak like everyone else. Raveloe's is a surprisingly homogeneous society, with no wide range of rank or wealth.

There is Squire Cass:

He was only one among several landed parishioners, but he
alone was honoured with the title of Squire; for though
Mr. Osgood's family was also understood to be of timeless
origin—the Raveloe imagination having never ventured
back to that fearful blank when there were no Osgoods—
still, he merely owned the farm he occupied; whereas
Squire Cass had a tenant or two, who complained of the
game to him quite as if he had been a lord.

The gentry of Raveloe are scarcely less naïve than the
rustics, their lives almost as confined and sequestered.

It is, however, in the character of Dolly Winthrop that
the folk-wisdom of Raveloe is fully realised as the truth
of feeling. Dolly is that rarest character in fiction, a positively good person. She is not in the least sentimentalised
but shown in her daily actions of charity, carried out almost at the unconscious level, since the doing of them is,
as it were, a natural thing. Here, she is a totally successful character as Dinah Morris, for instance, is not.

We see her with Marner on the Sunday morning before Christmas:

"Dear heart!" said Dolly, pausing before she spoke
again. "But what a pity it is you should work of a Sunday,
and not clean yourself—if you *didn't* go to church; for if
you'd a roasting bit, it might be as you couldn't leave it,
being a lone man. But there's the bakehus, if you could
make up your mind to spend a twopence on the oven now
and then—not every week, in course—I shouldn't like to do
that myself—you might carry your bit o' dinner there, for
it's nothing but right to have a bit o' summat hot of a
Sunday, and not to make it as you can't know your dinner
from Saturday. But now, upo' Christmas day, this blessed
Christmas as is ever coming, if you was to take your dinner
to the bakehus, and go to church, and see the holly and
the yew, and hear the anthim, and then take the sacramen',
you'd be a deal the better, and you'd know which end you
stood on, and you could put your trust in Them as knows

better nor we do, seeing you'd ha' done what it lies on us
all to do."

George Eliot tells us that Dolly made her exhortation
"in the soothing, persuasive tone with which she would
have tried to prevail on a sick man to take his medicine,
or a basin of gruel for which he had no appetite."

As one reads *Silas Marner,* one feels of George Eliot
(as one sometimes does of Hardy) that only a radical
free-thinker cut off from her (or his) roots could have
had quite so intense a nostalgia for the traditional past.

But besides the delineation of country life and humour
and of a world gone for ever, there is something else in
Silas Marner—the moral vision which is, in a sense, the
spine of the story. It is a myth of rebirth, but it is also a
novel of redemption. It has a double action, Marner's and
the young Squire Godfrey Cass's. Marner is the unwitting
agent of Cass's redemption, just as Cass's behaviour is
the unwitting cause of Marner's rebirth. But, as George
Eliot observed to Blackwood, "The Nemesis is a very
mild one," for all that the novel contains one of her most
intransigent statements of her moral view: "Favourable
chance . . . is the god of all men who follow their own
devices instead of obeying a law they believe in. . . . The
evil principle deprecated in that religion is the orderly
sequence by which the seed brings forth a crop after
its kind." In *Silas Marner,* certainly, the seed brings forth
the crop after its kind. Retribution falls on the guilty:
Cass is punished for his youthful sin through his daughter
Eppie's refusing to acknowledge him as her father and
cleaving to Marner. Yet, of the two facets of the novel,
that of rebirth and that of redemption, it is the first that
is dominant, and thereby the novel obtains much of its
power to move and to delight, for the myth of rebirth
takes us beyond morality.

5

"I REMEMBER MY WIFE telling me, at Witley," wrote Cross in *George Eliot's Life*, "how cruelly she had suffered at Dorking from working under a leaden weight at this time. The writing of 'Romola' ploughed into her more than any of her other books. She told me she could put her finger on it as marking a well-defined transition in her life. In her own words, 'I began it a young woman, —I finished it an old woman.' "

The strain she underwent in writing *Romola* suggests that the task she had set herself was unnatural to her; her struggles with *Romola* exist in vivid contrast to the ease and speed with which she wrote *Silas Marner*. But with that novel, the writing of which interrupted her progress with *Romola*, she was in her element, re-creating her own childhood memories, drawing upon the past in which she herself was rooted. With *Romola* she had no such help. Writing it was hard slogging all the way, a feat of research rather than creation; and the pity of it was that, in the end, it was hard slogging to little purpose. While it is impossible to read *Romola* without respect, it is also impossible to read it with much pleasure or more than once. It was a tremendous labour misconceived.

It was on August 28, 1860, that she wrote to Blackwood:

I think I must tell you the secret, though I am distrusting my power to make it grow into a published fact. When we were in Florence, I was rather fired with the idea of writing a historical romance—scene, Florence; period, the close of the fifteenth century, which was marked by Savonarola's career and martyrdom. Mr. Lewes has encouraged me to persevere in the project, saying that I should probably do something in historical romance rather different in character from what has been done before.

By the following January, however, she was writing *Marner,* "which came *across* my other plans by a sudden inspiration." Five weeks after finishing it, she set out with Lewes for a second visit to Florence, where they stayed "thirty-four days of precious time." "Will it all be in vain?" she asked herself in her journal. "Our morning hours were spent in looking at streets, buildings, and pictures, in hunting up old books, at shops or stalls, or in reading at the Magliabecchian Library." On October 7, she began the first chapter; by November 6, she was "so utterly dejected, that in walking with G. in the Park, I almost resolved to give up my Italian novel." However, by December 12, she had finished writing her plot, "of which I must make several other draughts before I begin to write my book."

At the end of the year, she recorded in her journal some of the books she had read during the second half of 1861. They included the *Histoire des ordres religieux,* Sismondi's *History of the Italian Republics,* Montalembert's *Monks of the West,* Savonarola's poems, Renan's *Études d'histoire religieuse,* Gibbon on the revival of Greek learning, Nardi's *History of Florence,* Burlamacchi's *Life of Savonarola,* Villari's *Life of Savonarola,* Pulci's *Morgante,* Mrs. Jameson's *Sacred and Legendary Art,* Politian's *Epistles,* Tiraboschi's *Storia della lèttteratura italiana,* Manni's *Life of Burchiello,* the works of Machiavelli, Petrarch's letters, and the letters of Filelfo, Lastri and Varchi. She had also frequented the British Museum to study fifteenth-century Italian costume. Then, on January 1, 1862, she underlined the following in her journal: "*I began again my novel of 'Romola'.*" She was, in fact, reading for the novel throughout the time she was writing it. Her journal entry for January 26, 1862, reads:

> Detained from writing by the necessity of gathering particulars: 1st, about Lorenzo de Medici's death; 2nd, about the possible retardation of Easter; 3rd, about Corpus Christi day; 4th, about Savonarola's preaching in the Quaresima of 1492. Finished "La Mandragola"—second time reading for the sake of Florentine expressions—and began "La Calandra."

On May 16, 1863, she could write: "Finished Part XIII. Killed Tito in great excitement," and on June 9, "Put the last stroke to 'Romola.' *Ebenezer!*" And by then the novel was already being published as a serial in *Cornhill;* she was paid £7,000 for it.

All that diligent research could do, she had done; as she told Frederic Harrison some years later: "I took unspeakable pains in preparing to write 'Romola'—neglecting nothing I could find that would help me to what I may call the 'idiom' of Florence, in the largest sense one could stretch the word to"; while to R. H. Hutton, on the book's publication, she wrote: "It is the habit of my imagination to strive after as full a vision of the medium in which a character moves as of the character itself. The psychological causes which prompted me to give such details of Florentine life and history as I have given, are precisely the same as those which determined me in giving the details of English village life in 'Silas Marner,' or the 'Dodson' life, out of which were developed the destinies of poor Tom and Maggie."

Yet, as the novel shows, research, however diligent, could not take the place of the instinctive knowledge in the blood and at the finger-tips that shaped Tom and

Maggie. Dolly Winthrop has more reality in her than all the characters in *Romola* put together. As Henry James said: "More than any of her novels it was evolved from her moral consciousness—a moral consciousness encircled by a prodigious amount of literary research," in lieu of the felt experience of men and women living in communities that went to the making of the rest of her fiction.

This is much more plain now, of course, than it was when the book first appeared, and one reason why this is so is that no novels date more quickly than historical novels. Historical novels reflect the times they are written in, and their assumptions, not the times they are written about. We need not be surprised at Trollope's enthusiasm for *Romola:*

> Adam Bede, Mrs. Poyser, and Marner have been very dear to me; but excellent as they are, I am now compelled to see that you can soar above even their heads. The description of Florence,—little bits of Florence down to a close nail, and great facts of Florence up to the very fury of life among those full living nobles,—are wonderful in their energy and in their accuracy. The character of Romola is artistically beautiful,—a picture exceeded by none that I know of any girl in any novel. It is the perfection of pen painting. . . .

The praise is generous and, in a sense, just. The facts of life in fifteenth-century Florence, George Eliot knew as well as anyone could:

> . . . its strange web of belief and unbelief; of Epicurean levity and fetichistic dread; of pedantic impossible ethics uttered by rote, and crude passions acted out with childish impulsiveness; of inclination towards a self-indulgent paganism, and inevitable subjection to that human conscience which, in the unrest of a new growth, was filling their air with strange prophecies and presentiments.

But the facts are given a Victorian twist; it is unlikely, for example, that a novelist writing today of fifteenth-century Florence would use such a phrase as "crude passions acted out with childish impulsiveness," since it would probably seem to him equally applicable to the twentieth century.

Yet *Romola* gives us facts enough, and to spare, about fifteenth-century Florence. What is lacking is the sense of its special quality of life, the quality that distinguished it from, for example, that of Victorian England. Mrs. Joan Bennett suggests that George Eliot was attracted to Savonarola's Florence because of its apparent similarities with Newman's England. What evidence there is for this, I do not know, but one can only say that any apparent similarities there are were dangerously misleading; or so it must seem today, when one feels that certain elements of Dreiser's novels *The Financier* and *The Titan,* studies of the nineteenth-century American scene though they are, are closer to aspects of Renaissance Italy than *Romola* is to any of them.

What one misses in *Romola* the whole time is imaginative insight into the past, the kind of insight, the flashes of intuition into alien modes of thought and feeling, that one finds in Scott, in, for example, the description in *Quentin Durward* of Louis XI's adoration of the images in his hat. Such a passage expands the reader's perceptions and therefore his sympathies. There is nothing like this, which is the real value and justification of historical fiction, in *Romola.* Instead, there is a vast amount of solid information about Renaissance Florence. To read *Romola* now is rather like going into a museum after hours: there is the city and the period, reconstructed and mounted in fascinating detail and with scrupulous care, and in the aisles between the exhibits and the artifacts, a Victorian fancy-dress masquerade is going on. This is apparent straightaway with the minor characters, such as Nello, the barber; they are speaking "lines" put into their mouths. It is just as apparent with the major figures Tito, Romola and Savonarola himself.

But with these, there is a difference, for they are the products, strikingly, of George Eliot's moral consciousness. That is where their interest lies, and what life they have derived from it. It is, at best, a shadowy life; they are characters, as it were, in the abstract. They are thoroughly and admirably analysed, but in the absence of roots in a reality the author has deeply felt and apprehended, analysis is not enough. Indeed, one feels all the time that the analysis never quite matches the actions they have to perform, which stem from their author's notions of fifteenth-century Florentine life. This matters least with Savonarola, whose presence is felt throughout the novel but who is all the same, seen as a character, a marginal figure. It is fatal, however, to Tito and Romola, both of whom appear as stalking-horses behind which George Eliot hunts down examples of moral failure and moral aspiration. They are thoroughly George Eliotish characters. Of the two, Tito is the more interesting and the more alive; inevitably, since, for the first half of the novel, Romola's is largely a passive role. She sums him up in the last pages of the novel:

"There was a man to whom I was very near, so that I could see a great deal of his life, who made almost every one fond of him, for he was young, and clever, and beautiful, and his manners to all were gentle and kind. I believe, when I first knew him, he never thought of anything cruel or base. But because he tried to slip away from everything that was unpleasant, and cared for nothing else so much as his own safety, he came at last to commit some of the basest deeds—such as make men infamous. He denied his father, and left him to misery; he betrayed every trust that was reposed in him, that he might keep himself safe and get rich and prosperous. Yet calamity overtook him."

Tito is doomed for the moment when he believes, because it is in his interests so to believe, that his father is dead:

> When, the next morning, Tito put this determination into action he had chosen his colour in the game, and had given an inevitable bent to his wishes. He had made it impossible that he should not from henceforth desire it to be the truth that his father was dead; impossible that he should not be tempted to baseness rather than that the precise facts of his conduct should not remain for ever concealed.

He is involved in a gradual but inexorable corruption:

> . . . our deeds are like children that are born to us; they live and act apart from our own will. Nay, children may be strangled, but deeds never: they have an indestructible life both in and out of our consciousness; and that dreadful vitality of deeds was pressing hard on Tito for the first time.

One can hardly not think that George Eliot presses hard on him as well:

> Tito was experiencing that inexorable law of human souls, that we prepare ourselves for sudden deeds by the reiterated choice of good or evil which gradually determines character.

Or again:

> . . . all the motives which might have made Tito shrink from the triple deceit that came before him as a tempting game, had been slowly strangled in him by the successive falsities of his life.
>
> Our lives make a moral tradition for our individual selves, as the life of mankind at large makes a moral tradition for the race; and to have once acted nobly seems a reason why we should always be noble. But Tito was feeling the effect

of an opposite tradition: he had won no memories of self-conquest and perfect faithfullness from which he could have had a sense of failing.

Thus analysed, Tito seems the George Eliot character isolated almost under laboratory conditions. The trouble is that the analysis and the language in which the analysis is couched have little to do with the Renaissance man whose behaviour is being described. There are moments, it is true, when the nineteenth century and fifteenth century fuse, as at the point when Tito challenges Romola:

> "At least," he added, in a slightly harder tone, "you will endeavour to base our intercourse on some other reasoning than that, because an evil deed is possible, I have done it. Am I alone to be beyond the pale of your extensive charity?"

Here, as Robert Speaight points out in his *George Eliot*, we have Grandcourt of *Daniel Deronda* to the life. At this moment, Tito is alive.

But these moments are rare, and even rarer when we are following the course of Romola's life. She is the quintessential George Eliot heroine, though a little nearer to Dorothea Brooke than to Maggie Tulliver. But she is faint by comparison because her life and its circumstances, the very problems of behaviour and faith that beset her, strike us as intellectual constructs; they do not compel our belief that this, in the terms of the miniature world the author is creating, is how things were and could only have been.

6

GEORGE ELIOT'S was an essentially nineteenth-century genius, an English nineteenth-century genius at that. One cannot easily imagine any period of history before her in the rendering of which her talents would have been at home and at their ease with the flexibility with which she exercised them upon the society of her own times and a little earlier—except possibly that of the English seventeenth century and the struggles between Puritan and Cavalier, between which, one feels, her sympathies might have been fruitfully divided. Certainly it seems that a story of the Italian Renaissance was the worst possible mould in which she could have let her genius flow. Yet though *Romola* may now appear, for the most part, so much deadwood and though, seeing what was the result, one may regret the time and the labour she spent on it and the agony she endured while writing it, the effort was not by any means all waste. It had its reward, for herself as novelist and for us as readers, in the novels that followed—so much so that *Romola* divides her work as a kind of water-shed, and what she learned in writing *Romola* is apparent immediately when we turn to her next novel, *Felix Holt, the Radical.*

Writing *Romola,* she had been outside her material in a way that she had not been in the novels that had gone before. In *Felix Holt* she returned in part to what was familiar to her, to what indeed was the stuff of memory, but how differently she handles it, compared with what she had done in, for example, *The Mill on the Floss.* In *Felix Holt* she has it, by comparison, at arm's length; she is, one might say, the master of her memories instead of being mastered by them. She has become much more impersonal; she has gained enormously in objectivity; and because she sees more clearly, her dramatic sense is so much the more powerful. She can present big scenes directly; and when she comments, her comments are the more authoritative because the more terse. Her intellect, one feels, has itself become creative; and as a consequence, the range of her artistic sympathy, seen in the range of characters she can render, is greatly widened.

All this is plain in the brilliant opening chapters of *Felix Holt* describing the meeting, after many years, of Mrs. Transome and her son Harold. It is plain, for instance, in the following:

As Mrs. Transome descended the stone staircase in her old black velvet and point, her appearance justified Denner's personal compliment. She had that high-born imperious air which would have marked her as an object of hatred and reviling by a revolutionary mob. Her person was too typical of social distinctions to be passed by with indifference by any one: it would have fitted an empress in her own right, who had had to rule in spite of faction, to dare the violation of treaties and dread retributive invasions, to grasp after new territories, to be defiant in desperate circumstances, and to feel a woman's hunger of the heart for ever unsatisfied. Yet Mrs. Transome's cares and occupations had not been at all of an imperial sort. For thirty years she had led the monotonous narrowing life which used to be the lot of our poorer gentry, who never went to town, and were probably not on speaking terms with two out of the five families whose parks lay within the distance of a drive. . . . Mrs. Transome, whose imperious will had availed little to ward off the great evils of her life, found the opiate for her discontent in the exertion of her will about smaller things. She was not cruel, and could not enjoy thoroughly what she called the old woman's pleasure of tormenting; but she liked every little sign of power her lot had left her. She liked that a tenant should stand bareheaded below her as she sat on horseback. She liked to insist that work done without her orders should be undone from beginning to end. She liked to be curtsied and bowed to by all the congregation as she walked up the little barn of a church. She liked to change a labourer's medicine fetched from the doctor, and substitute a prescription of her own. If she had only been more haggard and less majestic, those who had glimpses of her outward life might have said she was a tyrannical, griping harridan, with a tongue like a razor. No one said exactly that; but they never said anything like the full truth about her, or divined what was hidden under that outward life—a woman's keen sensibility and dread, which lay screened behind all her petty habits and narrow notions, as some quivering thing with eyes and throbbing heart may lie crouching behind withered rubbish. The sensibility and dread had palpitated all the faster in the prospect of her son's return; and now that she had seen him, she said to herself, in her bitter way, "It is a lucky eel that escapes skinning. The best happiness I shall ever know, will be to escape the worst misery."

Admittedly, power of this order is intermittent in *Felix Holt*, which cannot be judged a successful novel. In some ways, it anticipates *Daniel Deronda* in pursuing two different themes linked together by a clumsy plot. One theme, that suggested in the title of the novel, is moral, political and exhortatory: the inadequacy of universal suffrage as an end in itself divorced from the will to use it wisely; the other is George Eliot's perennial theme of retribution that unfailingly follows wrong action. For her setting, she went back to the agitations that attended the passing of the first Reform Act of 1832 as she had witnessed them herself as a girl in Nuneaton.

She began the novel on March 29, 1865, and finished it on May 31, 1866, "in a state of nervous excitement that had been making my head throb and my heart palpitate all the week before." Writing it, she went through her usual agonies, "going doggedly to work . . . seeing what determination can do in the face of despair"; but that she wrote it so quickly shows how much more congenial it was to her than *Romola*. She did her stint of research. As she wrote to Blackwood a month before writing the last lines:

I took a great deal of pains to get a true idea of the period. My own recollections of it are childish, and of course, disjointed, but they help to illuminate my reading. I went through the "Times" of 1832-33 at the British Museum, to be sure of as many details as I could. It is amazing what strong language was used in those days, especially about the Church. "Bloated pluralists," "Stall-fed dignitaries," etc., are the sort of phrases conspicuous.

For legal information essential to her plot, she leaned on Frederic Harrison, the Positivist lawyer.

Next to *Romola*, *Felix Holt, the Radical* has been the least regarded of George Eliot's novels, and very largely because of the plot, which turns on points of the law relating to inheritance that have baffled most readers and were, indeed, challenged by some among the knowledgeable when the book first appeared. The plot could not be more cumbersome and practically defies synopsis. It involves mysteries of parenthood, which are common enough in Victorian fiction and, indeed, in George Eliot's own fiction; it forces her into the exploitation of coincidental encounters between characters altogether too flagrant to be acceptable. It obtrudes, and in a purely mechanical way.

One can, of course, always justify the mechanical plot
of missing heirs, mysterious paternity, suppressed wills,
lost documents and the rest that came down to the Victorians from Fielding, who got them from the theatre: at
least they enable those who use them to bring into the
action whole sets of contrasted social scenes. This is
much, and one can see in our own time a novelist like
Angus Wilson in *Anglo-Saxon Attitudes* seeking for the
equivalent of the Victorian plot simply in order to
enable him to render a wide canvas of characters from
disparate walks of life. Yet experience shows that almost
nobody remembers plots of this kind, creaking construc-
tions as they are; and if remembered, they can only
seem grotesquely irrelevant to the real life of the novels
that incorporate them. This is certainly so of *Felix
Holt*. George Eliot did her earnest best to relate everything in the novel to everything else; but it remains true
that, however one considers it, it falls into two parts, each
with its own centre of interest, which touch and at
times intermingle—in defiance, it seems in memory, of the
demands of the plot.

The two parts coincide with the two themes; one may
be called the Felix Holt theme, the other the Mrs.
Transome theme. The second contains some of the
finest work George Eliot ever did; the first has its moments of interest and is certainly not nearly as bad as
some critics have maintained. In it George Eliot may be
said to go back to the town scenes of *Silas Marner*, though
they are brought forward, made more modern, by a
generation. The subject is dissent, religious and political,
the latter springing out of the former. The representative
of religious dissent, the Reverend Mr. Lyon, is rendered
in George Eliot's earlier mode, in terms of affectionate
irony. The affection is perhaps a little too indulgent;
the humour and the pathos implicit in this innocent, good
old man are a little too easy. Still, Mr. Lyon gives rise to

much pleasant comedy, as may be seen in this passage in which he is addressing his servant:

> "Lyddy," said Mr. Lyon, falling at once into a quiet conversational tone, "if you are wrestling with the enemy, let me refer you to Ezekiel the thirteenth and twenty-second, and beg of you not to groan. It is a stumbling-block and offence to my daughter; she would take no broth yesterday, because she said you had cried into it. Thus you cause the truth to be lightly spoken of, and make the enemy rejoice. If your face-ache gives him an advantage, take a little warm ale with your meat—I do not grudge the money."

And the comic implications in the conflict between dissent and the established church are nicely drawn out.

Felix Holt is a character more interesting potentially than in fact. The conception of the educated, class-conscious young man who will not leave his class, for all his advantages and opportunities, is a striking one; and his milieu is admirably drawn. His uncomprehending widowed mother, with her touching faith in her husband's patent medicines, which Felix scorns as quackery, is an entertaining minor character. But Felix, as is the way of the heroes George Eliot admires, is too good to be true. He habitually speaks as no man ever did, addressing Esther always as though she were a public meeting or a class of schoolgirls to be scolded. What he says is wise, but his very wisdom is a hindrance to any quality of conviction he should carry: he would, one feels, have been more convincing if George Eliot had made him a revolutionary instead of a radical who had made himself drunk on Carlyle. But he is the puppet of his author's views, and it would have been impossible for her to have made him a revolutionary.

His woodenness is such that it is not easy to see how Esther Lyon could fall in love with him, the more especially since the realisation that she loves him is tantamount to a kind of conversion. As we first meet her, she is a brilliant, selfish, witty girl, as much a princess in exile as Gwendolen Harleth in *Daniel Deronda;* indeed, one might call her a Gwendolen Harleth of the dissenting chapel. As such, she is admirably done, and—something new in George Eliot—not at all, as in the earlier novels, the heroine with whom her author identifies herself.

In the end, against all odds, the conversion of the young woman who aspires after goodness and the willing subjection to the superior being seems to me convincing—but not because of Felix Holt. Its truth is registered when, as the possible rightful heir of the Transome estates, she goes to stay at Transome Court and is wooed by Harold Transome. She is in a world of luxury she had never known before.

> And yet, this life at Transome Court was *not* the life of her day-dreams: there was dullness already in its ease, and in the absence of high demand; and there was a vague consciousness that the love of this not unfascinating man who hovered about her gave an air of moral mediocrity to all her prospects. She would not have been able perhaps to define this impression; but somehow or other by this elevation of fortune it seemed that the higher ambition which had begun to spring in her was for ever nullified. All life seemed cheapened; as it might seem to a young student who, having believed that to gain a certain degree he must write a thesis in which he would bring his powers to bear with memorable effect, suddenly ascertained that no thesis was expected, but the sum (in English money) of twenty-seven pounds ten shillings and sixpence.

Esther's conversion rings true because of the absolute reality of Transome. He is most brilliantly rendered, as indeed is all of what might be called the Transome section of the novel, the lives of the gentry and their servants. They are depicted, one feels, with utter certainty, Sir Maximus Debarry, the Reverend John Lingon and Mrs. Transome's maid Denner alike. But it is with Transome, his mother and the lawyer Jermyn that George Eliot achieves triumphs beyond anything she had done before.

Transome, in terms of his creator's work, belongs to the company of Lydgate and Grandcourt. That is to say, he impresses by his masculinity. These men of George Eliot's are very much men as seen by a woman, but this does not in the least diminish their truth. They are not figures of wish fulfilment, like Charlotte Brontë's Rochester. One feels that years of experience of life, years of observation of men in society, have gone to their making. They are men seen, as it were, as the opposite of women, the masculine as opposed to the feminine. They are hard and arrogant; their forte is action; as we have seen, for a brief moment in *Romola*, Tito anticipates them. All of them suffer, because of their masculinity, their unconscious strength as lords of creation, from "spots of commonness," to borrow the phrase George Eliot applies to Lydgate. Her criticism of them, indeed, is continuous and explicit; we are left in no doubt about how we are expected to see them; yet, for all this, they are presented with remarkable objectivity. For what is not in question at any time is their quality of formidableness.

Because of their exclusive masculinity and the assumptions that belong to it, they seem born to make women, whether as wives or mothers, suffer; and there can be few passages in fiction more moving than the encounters, with which *Felix Holt* begins, between Mrs. Transome and Harold, between the proud woman and the heedless son who has returned, having made a fortune in the Levant, with political ambitions in directions quite other than anything she has visualised. But no less good are the scenes towards the end of the novel in which Harold discovers the facts of his paternity, that he is the son of the attorney Jermyn, whom he has treated with gentlemanly contempt. His attitude to Jermyn is, indeed, very much that of his uncle the Reverend John Lingon, who describes the lawyer as "A fat-handed, glib-tongued fellow, with a scented cambric handkerchief; one of your educated low-bred fellows; a foundling who got his Latin for nothing at Christ's Hospital; one of your middle-class upstarts who want to rank with gentlemen, and think they'll do it with kid gloves and new furniture." Jermyn, George Eliot tells us,

> . . . chose always to dress in black, and was especially addicted to black satin waistcoats, which carried out the general sleekness of his appearance; and this, together with his white, fat, but beautifully shaped hands, which he was in the habit of rubbing gently on his entrance into a room, gave him very much the air of a lady's physician. Harold remembered with some amusement his uncle's dislike of those conspicuous hands; but as his own were soft and dimpled, and as he too was given to the innocent practice of rubbing those members, his suspicions were not yet deepened.

Jermyn, for Harold, is simply a rascal who has battened on his mother for years and must be punished accordingly. As George Eliot says: "Moral vulgarity cleaved to him like an hereditary odour." Hence the bitterness of the final confrontation and the discovery:

> Jermyn walked quickly and quietly close up to him. The two men were of the same height, and before Harold looked round Jermyn's voice was saying, close to his ear, not in a whisper, but in a hard, incisive, disrespectful and yet not loud tone,—
> "Mr. Transome, I must speak to you in private."
> The sound jarred through Harold with a sensation all the more insufferable because of the revulsion from the satisfied, almost elated, state in which it had seized him. He started and looked round into Jermyn's eyes. For an instant, which seemed long, there was no sound between

them, but only angry hatred gathering in the two faces. Harold felt himself going to crush this insolence: Jermyn felt that he had words within him that were fangs to clutch this insolent strength, and bring forth the blood and compel submission. And Jermyn's impulse was the more urgent. He said, in a tone that was rather lower, but yet harder and more biting—

"You will repent else—for your mother's sake."

At that sound, quick as a leaping flame, Harold had struck Jermyn across the face with his whip. The brim of the hat had been a defence. Jermyn, a powerful man, had instantly thrust out his hand and clutched Harold hard by the clothes just below the throat, pushing him slowly so as to make him stagger.

By this time everybody's attention had been called to this end of the room, but both Jermyn and Harold were beyond being arrested by any consciousness of spectators.

"Let me go, you scoundrel!" said Harold, fiercely, "or I will be the death of you."

"Do," said Jermyn in a grating voice; "*I am your father.*"

In the thrust by which Harold had been made to stagger backwards a little, the two men had got very near the long mirror. They were both white; both had anger and hatred in their faces; the hands of both were upraised. As Harold heard the last terrible words he started at a leaping throb that went through him, and in the start turned his eyes away from Jermyn's face. He turned them on the same face in the glass with his own beside it, and saw the hated fatherhood reasserted.

Yet in this suddenly revealed relationship, the tragic figure is that of his mother:

"*He* has said—said it before others—that *he* is my father."

He looked still at his mother. She seemed as if age were striking her with a sudden wand—as if her trembling face were getting haggard before him. She was mute. But her eyes had not fallen; they looked up in helpless misery at her son.

Her son turned away his eyes from her, and left her. In that moment Harold felt hard: he could show no pity. All the pride of his nature rebelled against his sonship.

Mrs. Transome is probably George Eliot's finest and most intense embodiment of retribution for wrong action. She has sinned long before the novel begins and stands before us immediately and with startling dramatic power as the representation of the wasted life, wasted because committed to a sin which cannot be undone or the consequences of which evaded. The awareness of it shapes her whole being, and secret remorse eats her away like a cancer. All George Eliot's most passionate apprehension of the workings of sin go into her making; she is, not excepting Gwendolen Harleth, her creator's most powerful symbol of the bleakness and bitterness of despair and guilt that consume their possessor.

7

IN MIDDLEMARCH, George Eliot takes up and develops one side of *Felix Holt:* the lives of the gentry and their relations with the professional, commercial and political life of the town that is the local metropolis. The period, too, is much the same, the months immediately preceding the passing of the Reform Act of 1832. Though it is not apparent as we read it, the novel changed in the course of writing; it is indeed in a sense a fusion of two novels. In her journal entry for January 1, 1869, George Eliot wrote:

> I have set myself many tasks for the year. I wonder how many will be accomplished?—a novel called "Middlemarch," a long poem on Timoleon, and several minor poems.

But this novel called *Middlemarch* was, at that stage, much smaller and narrower in scope than the one we know now. It was to be a study of provincial life, with its central character a doctor; the Vincys were also there from the beginning. It went slowly, and its progress was attended with all the usual doubts that assailed George Eliot while working on a novel: "I do not feel," she wrote in September, "very confident that I can make anything very satisfactory of 'Middlemarch.' I have need to remember that other things which have been accomplished by me were begun under the same cloud." A day before, she had commented in her journal: "I have achieved little during the last week, except reading on medical subjects—Encyclopaedia about the Medical Colleges, 'Cullen's Life,' Russell's 'Heroes of Medicine,' etc."

After that, *Middlemarch* seems to have dragged. Then at the beginning of December, 1870, she notes: "I am experimenting in a story ('Miss Brooke') which I began without any very serious intention of carrying it out lengthily. It is a subject which has been recorded among my possible themes ever since I began to write fiction, but will probably take new shapes in the development." By the last day of the year, she had written a hundred pages—"good printed pages"—of *Miss Brooke*. That is, in fact, the last we hear of *Miss Brooke;* when we next hear of "my novel," it has obviously merged into the *Middlemarch* which is now impossible to imagine without it.

The novel, the longest George Eliot wrote, appeared serially in eight parts, the last of which was published in December, 1873. George Eliot wrote in her journal that "no former book of mine has been received with more enthusiasm—not even 'Adam Bede.' "

Never was enthusiasm more justified. She had produced something that was quite new in English fiction, a panorama of English social life on a scale unattempted by anyone before. Not only is it the most comprehensive of her novels, it is also the most successful artistically, if we set aside *Silas Marner*, which is of so much smaller a range as scarcely to be comparable. The characters and relationships between characters that she achieves in *Middlemarch* may not be better than the best things in *Felix Holt* and *Daniel Deronda*, but *Middlemarch* exists as a whole as those novels do not. One does not have to apologise for and shrug off whole sections as one does with the others.

Middlemarch is unique in George Eliot's work in being a beautiful composition, and this despite the vast mass of material with which George Eliot was working. There are no less than four major centres of interest to the novel; one might, indeed, say four major plots: Dorothea Brooke's story, the story of Lydgate and his marriage, the story of the Garths and the fall of the banker Bulstrode. George Eliot brings them into relationship one with another without apparent strain, so that they interlock and, doing so, support one another. Each has its own satellite centres of interest, and together they compose a network of relationships enclosing the whole life and movements of events and opinion in a provincial city and the country around it.

Of these four major stories, however, two are more significant than the others—Dorothea's and Lydgate's, partly because of the especial magnetism of these two characters and partly because it is in them that the life of the novel is at its intensest and most deeply felt. And of these two stories it is difficult not to give priority to Dorothea's, since Dorothea herself is particularly close to her creator and, even more than Maggie Tulliver, is the quintessential George Eliot heroine. She is, as it were, Maggie Tulliver heightened by social position, wealth and a comparative independence from the ties of family, all of which give her a sublime faith in herself and the rightness of her behaviour that Maggie lacks.

This may perhaps be put in a different way. Maggie has at least one thing in common with Esther Lyon and Gwendolen Harleth: she is a princess in exile, and the sense of being in exile goes to the heart of her situation. Dorothea is not a princess in exile; quite the opposite: she is, so to say, a reigning princess, both in her own eyes and in those of the characters about her. In this respect, she is nearer to James's Isabel Archer and Milly Theale than to Maggie and the other Eliot heroines.

She is a great conception, conceived and created with remarkable intransigence:

Miss Brooke had that kind of beauty which seems to be
thrown into relief by poor dress. Her hand and wrist were
so finely formed that she could wear sleeves not less bare
in style than those in which the Blessed Virgin appeared
to Italian painters; and her profile as well as her stature
and bearing seemed to gain the more dignity from her
plain garments, which by the side of provincial fashion
gave her the impressiveness of a fine quotation from the
Bible—or from one of our elder poets—in a paragraph in
today's newspaper.

It is with these sentences that the novel opens; no
wonder we feel that Dorothea dominates the book.
George Eliot goes on, a page or so later:

She could not reconcile the anxieties of a spiritual life
involving eternal consequences, with a keen interest in
guimp and artificial protusions of drapery. Her mind was
theoretic, and yearned by its nature after some lofty conception of the world which might frankly include the parish
of Tipton and her own rule of conduct there; she was
enamoured of intensity and greatness, and rash in em-
bracing whatever seemed to her to have these aspects;
likely to seek martyrdom, to make retractions, and then to
incur martyrdom after all in a quarter where she had not
sought it.

Again, following Casaubon's letter proposing marriage:

All Dorothea's passion was transfused through a mind strug-
gling towards an ideal life; the radiance of her transfigured
girlhood fell on the first object that came within its level.
The impetus with which inclination became resolution was
heightened by those little events of the day which had
roused her discontent with the actual conditions of her life.

The first passage establishes Dorothea's fineness and
austerity, what could be called her distinction of soul.
One might indeed feel that George Eliot is pitching her
claims for the girl too high. The other two quotations,

which are representative and could be replaced by any two of a score of others, continue to emphasize her spiritual aspirations, which amount almost to a lust after goodness and the life of self-sacrifice; nevertheless, their tone is critical. Dorothea is the victim of her own head-long spiritual enthusiasms, of her ignorance of the world and of herself, of her lack of self-discipline. Her be-haviour is almost that of a spoilt child who has been allowed her own way too long. Yet this in no way detracts from the genuine quality of Dorothea's spiritual ardour; we are not concerned simply with a girl who is deceived about herself. Dorothea is a much more complex being than, say, Jane Austen's Emma, and George Eliot's treat-ment of her is accordingly more subtle. In the prelude, the implicit comparison between her and St. Theresa has been made:

> That Spanish woman who lived three hundred years ago, was certainly not the last of her kind. Many Theresas have been born who have found for themselves no epic life wherein there was a constant unfolding of far-resonant ac-tion; perhaps only a life of mistakes, the offspring of a certain spiritual grandeur ill-matched with the meanness of opportunity.

"A certain spiritual grandeur ill-matched with the meanness of opportunity": that is the point. Dorothea is a St. Theresa, but a St. Theresa born at the wrong place and at the wrong time. And this leads us to what is George Eliot's great subject in *Middlemarch:* she is investigating human aspirations, in particular the aspirations to serve and to do good, in the light of the two factors that can make or mar their realisation. One is personal, the quali-ties within the human being himself—self-knowledge or lack of it, strength of will or, otherwise, human frailty; the other is social, the limits to action imposed upon the

individual by the society in which he is born, which
also dictates the channels open to aspiration.

All the chief characters in *Middlemarch*—Dorothea,
Lydgate, the Garths, Bulstrode—can be interpreted in
these terms; and when we do so, we see another reason
for the evident superiority of Dorothea and Lydgate.
They are, quite simply, superior as human beings. The
Garths are good; they know their duty and do it; their
goodness, in a sense, is an aspect of their humility. But
Mary Garth is no St. Theresa, and her father, conscientious craftsman though he is, is no dedicated scientist.
The very superiority of Dorothea and Lydgate, the loftiness of their aspirations, places them in a dimension of
tragedy—potential, if not realised in actuality—forbidden
to the Garths.

George Eliot criticises them throughout the novel as
she does not the Garths. The Garths' goodness is of a
kind that is beyond criticism; it is the less interesting
because of that. The most effective criticisms of Dorothea,
however, are not George Eliot's own comments and
analyses, masterly as these often are, but lie in her own
behaviour, her own speech and in what is felt and said
about her by those in her immediate circle. In the wonderful early scene, for instance, between Dorothea and
Celia after they have left the dinner-table following Mr.
Casaubon's first visit to Tipton Grange:

> "How very ugly Mr. Casaubon is!"
>
> "Celia! he is one of the most distinguished-looking men
> I ever saw. He is remarkably like the portrait of Locke. He
> has the same deep eye-sockets."
>
> "Had Locke those two white moles with hairs on them?"
>
> "Oh, I daresay! when people of a certain sort looked at
> him," said Dorothea walking away a little.
>
> "Mr. Casaubon is so sallow."
>
> "All the better. I suppose you admire a man with the
> complexion of a *cochon de lait*."

"Dodo!" exclaimed Celia, looking after her in surprise. "I never heard you make such a comparison before."

"Why should I make it before the occasion came? It is a good comparison: the match is perfect."

Miss Brooke was clearly forgetting herself, and Celia thought so.

"I wonder you show temper, Dorothea."

"It is so painful in you, Celia, that you will look at human beings as if they were merely animals with a toilette, and never see the great soul in a man's face."

"Has Mr. Casaubon a great soul?" Celia was not without a touch of naïve malice.

"Yes, I believe he has," said Dorothea, with the full voice of decision. "Everything I see in him corresponds to his pamphlet on Biblical Cosmology."

"He talks very little," said Celia.

"There is no one for him to talk to."

That is great comic writing, but its immediate purpose is to point out the fact that, superior though she is to all about her, Dorothea is blind as lesser beings are not. The symbol of her blindness is her infatuation—a spiritual infatuation—with Casaubon: " 'She says, he is a great soul.—A great bladder for dried peas to rattle in!' said Mrs. Cadwallader." Mrs. Cadwallader is one of George Eliot's most surely drawn minor characters; her estimate of Cassaubon is not quite just, but it is much nearer to the truth than Dorothea's self-deceiving notion of him, which is a creation of her own imagination. " 'He thinks with me,' said Dorothea to herself, 'or rather, he thinks a whole world of which my thought is but a poor two-penny mirror. And his feelings too, his whole experience —what a lake compared with my little pool!' " She is, of course, entirely wrong about him. He is an empty shell, a hollow man.

Yet with what lucid and beautiful charity George Eliot renders this withered, chilling figure; and with what skill, too, she associates him with, almost symbolises him in,

the gloomy attributes of his house, the Grange. Owner
and house are in a sense one:

> The building, of greenish stone, was in the old English
> style, not ugly, but small-windowed and melancholy-look-
> ing: the sort of house that must have children, many
> flowers, open windows, and little vistas of bright things, to
> make it seem a joyous home. In this latter end of autumn,
> with a sparse remnant of yellow leaves falling slowly
> athwart the dark evergreens in a stillness without sunshine,
> the house too had an air of autumnal decline, and Mr.
> Casaubon, when he presented himself, had no bloom that
> could be thrown into relief by that background.

The house, permanently sunless, enclosed by dark
evergreens, can stand for Casaubon's mind, which is
absurdly other than anything Dorothea has dreamed of:

> How was it that in the weeks since her marriage, Dorothea
> had not distinctly observed but felt with a stifling depres-
> sion, that the large vistas and wide fresh air which she had
> dreamed of finding in her husband's mind were replaced
> by anteroom and winding passages which seemed to lead
> nowhither?

Her disillusionment is inevitable and bitter. Yet Casaubon is not contemptible: he too exists in a dimension of
tragedy, and through George Eliot's comprehension of
him and pity for him, he is on the grand scale. She alone
could have conceived him, and never has desiccated
pedantry been more unerringly rendered; it is there in
every speech he utters, indeed in his solitary love-letter.
He is, as Sir James Chettam says, "a sort of parchment
code"; and even when most under Dorothea's spell can
achieve no better than "a smile like pale wintry sun-
shine." But the pedantry itself is not so important. Much
more important is what it hides, and though his end is
not tragic as hers is, the character of all George Eliot's he
is closest to is Mrs. Transome. Like her, he is living a lie

—the lie of his great book—knowledge of which haunts him ceaselessly. His very jealousy of Dorothea springs out of his fear that she has penetrated his secret, which is that his great book, which in the eyes of the world— and Dorothea's, too—is almost the symbol of his being, certainly his justification, is worthless. He is a terrifying figure of futility haunted by consciousness of itself; and reading him, what impresses us and moves us to pity is the sense of waste that informs his being. He too—and this is why we respond to him as we do—is potentially a St. Theresa figure: he too has a "certain spiritual grandeur ill-matched with the meanness of opportunity."

One could not easily see Lydgate as a St. Theresa figure. He is a much more modern type, the scientist; and he is essentially a man of his time:

> . . . about 1829 the dark territories of Pathology were a fine America for a spirited young adventurer. Lydgate was ambitious above all to contribute towards enlarging the scientific, rational basis of his profession. The more he became interested in special questions of disease, such as the nature of fever or fevers, the more keenly he felt the need for that fundamental knowledge of structure which just at the beginning of the century had been illuminated by the brief and glorious career of Bichat, who died when he was only one-and-thirty, but, like another Alexander, left a realm large enough for many heirs.

One of these Lydgate intends to be.

> He went to study in Paris with the determination that when he came home again he would settle in some provincial town as a general practitioner, and resist the irrational severance between medical and surgical knowledge in the interest of his own scientific pursuits, as well as of the general advance: he would keep away from the range of London intrigues, jealousies, and social truckling, and win celebrity, however slowly, as Jenner had done, by the independent value of his work.

His plan for his future, returning from Paris, is "to do good small work for Middlemarch, and great work for the world."

Lydgate, as one knows him in the novel, is the one character capable of meeting Dorothea on anything like her own level; there is, of course, also Ladislaw, but he is a pale shadow compared with Lydgate. Yet, partly because of the very largeness of his ambitions, George Eliot subjects Lydgate to a constant criticism as unrelenting as that she applies to Dorothea. It takes a very different form. It is George Eliot's criticism of the exclusively masculine, as we find it not only in Lydgate but also in Harold Transome and Grandcourt. There is, however, this distinction to be made between Lydgate and the other two: Lydgate's aspirations are such as to be in the end irreconcilable with the exclusively masculine.

George Eliot's criticism of Lydgate is minute and particular. His ambition is one thing, and by gaining the support of Bulstrode, who is building a new hospital, he seems in a fair way to achieving it. He is, in fact, defeated by the very circumstances of the provincialism he has elected to settle in. To this extent, his medical ambitions are ill-matched with the meanness of opportunity. But he is also in part defeated by himself. George Eliot says:

> . . . in the multitude of middle-aged men who go about their vocations in a daily course determined for them much in the same way as the tie of their cravats, there is always a good number who once meant to shape their own deeds and alter the world a little. The story of their coming to be shapen after the average and fit to be packed by the gross, is hardly ever told even in their consciousness; for perhaps their ardour in general unpaid toil cooled as imperceptibly as the ardour of other youthful loves, till one day their earlier self walked like a ghost in its old home and made

the new furniture ghastly. Nothing in the world more subtle
than the process of their gradual change! . . .

Lydgate did not mean to be one of those failures, and
there was the better hope of him because his scientific
interest soon took the form of a professional enthusiasm. . . .

All the same, Lydgate is of this number because of
what George Eliot calls his "spots of commonness." She
describes them in these terms:

> Lydgate's spots of commonness lay in the complexion of
> his prejudices which, in spite of noble intention and sym-
> pathy, were half of them such as are found in ordinary men
> of the world: that distinction of mind which belonged to
> his intellectual ardour, did not penetrate his feeling and
> judgment about furniture, or women, or the desirability of
> its being known (without his telling) that he was better
> born than other country surgeons. He did not mean to think
> much of furniture at present; but whenever he did so it
> was to be feared that neither biology nor schemes of reform
> would lift him above the vulgarity of feeling that there
> would be an incompatibility in his furniture not being of
> the best.

There is, in other words, a lack of balance, an uneven-
ness, in Lydgate's development. The independence of
mind and the sceptical, questing intellect, resolved to
take nothing on trust and to accept only what can be
proved by research and experiment, which distinguish
him as a scientist, have little part in his larger life as a
man. There, his values are purely conventional; he lacks
discrimination. The sign of this is his marriage to Rosamund Vincy, whom he sees, and who very largely sees
herself, as closely akin to the warrior's plaything.

George Eliot had little love for pretty girls as such;
indeed, she saw prettiness, sexual attractiveness as quali-
ties dangerous to the moral nature of those who possess
them. This is evident in the character of Rosamund, who

is a superb image of the self-satisfied acceptance of a totally conventional notion of social behaviour. She is brilliantly done, the pretty, vain girl whose upbringing has been rather better, in the snobbish sense, than her parents'. She knows, with all the complacency of a spoilt, pretty girl with a rich father, what is due her:

> "Mamma," said Rosamund, "when Fred comes down I wish you would not let him have red herrings. I cannot bear the smell of them all over the house at this hour of the morning."
>
> "Oh, my dear, you are so hard on your brothers! It is the only fault I have to find with you. You are the sweetest temper in the world, but you are so tetchy with your brothers."
>
> "Not tetchy, mamma: you never hear me speak in an unladylike way."

Lydgate is an obvious target for the exercise of her charms:

> . . . here was Mr. Lydgate suddenly corresponding to her ideal, being altogether foreign to Middlemarch, carrying a certain air of distinction congruous with good family, and possessing connections which offered vistas of that middle-class heaven, rank: a man of talent, also, whom it would be especially delightful to enslave: in fact, a man who had touched her nature quite newly, and brought a vivid interest into her life which was better than any fancied "might-be" such as she was in the habit of opposing to the actual.

Rosamund, then, and her appeal to him, embodies Lydgate's "spots of commonness." She has no understanding of the nature of his work and ambitions and, indeed, cannot have; her values are social in the narrowest of snobbish senses; and their marriage can be satisfactory only for so long as all is fair weather. It has to be emphasised, again, that the frustration of Lydgate's ambitions is not by any means entirely due to the "spots of commonness" of which his marriage is the symbol. It comes about primarily from the facts of life in Middlemarch itself, with its professional jealousies and intrigues, its political oppositions, the narrowness of its view of things; above all, perhaps, from his association with Bulstrode. To a very large degree, in other words, it comes about from the nature of things. It is Lydgate's tragedy that life with the demanding, self-blinded Rosamund cripples him in his efforts to withstand the forces of circumstance working against him.

The interest of the two other main stories in *Middlemarch* is much less compelling than those of Dorothea Brooke and Lydgate. For the Garths, George Eliot has only uncritical admiration; we see them from one stance always and see only one face of them. As a foil to Rosamund, Mary Garth is a failure; her liveliness does not compensate for George Eliot's inability to render her in depth. She suffers, like her parents, from the monotony of goodness. Her mother, it seems to me, is hardly "there" at all, except as a bundle of admired attributes; and her father is all too predictable both in action and in speech.

The failure with Bulstrode is of a different kind and is, indeed, only partial. Bulstrode is a most striking representation of Evangelicalism in one of its aspects: its relation to capitalism and economic and social power; and, as we first meet him in the novel, his formidableness is most formidably rendered. The failure lies in the crime remorse for which eats at his heart. It is altogether too melodramatic for the kind of novel George Eliot is writing; quite suddenly we find in a corner of what is mainly a modern novel a Victorian plot, with all the traditional apparatus of the Victorian plot—missing heirs, clumsy coincidental meetings and the rest.

There is one other failure in *Middlemarch*—the character of Will Ladislaw. One of his functions is like that

of Philip Wakem in *The Mill on the Floss:* he is to stand
for the free spirit, the larger world outside the action. He
is, where Dorothea is concerned, the voice of George
Eliot herself, as when he tells Dorothea:

> "The best piety is to enjoy—when you can. You are doing
> the most then to save the earth's character as an agreeable
> planet. And enjoyment radiates. It is of no use to try and
> take care of all the world; that is being taken care of when
> you feel delight—in art or in anything else. Would you turn
> all the youth of the world into a tragic chorus, wailing and
> moralising over misery? . . . You talk as if you had never
> known any youth. It is monstrous—as if you had had a
> vision of Hades in your childhood, like the boy in the
> legend. You have been brought up in some of those horrible
> notions that choose the sweetest woman to devour—like
> Minotaurs. And now you will go and be shut up in that
> stone prison at Lowick: you will be buried alive."

One applauds—and observes, too, that the note sounded
is one not often heard in George Eliot's fiction. But apart
from his role of author's spokesman, the impression Ladislaw makes is faint and somewhat puzzling. As an incar-
nation of his own belief that "the best piety is to enjoy,"
he is clearly not very satisfactory. George Eliot has no
superior as the creator—in a spirit of criticism and, one
can scarcely help feeling, of something like sexual hos-
tility—of male characters of the kind I have called
exclusively masculine, of which Lydgate is the exemplar
in *Middlemarch.* With men she obviously approves of
much more, she is much less successful, and the problem
of Ladislaw is akin to that of Stephen Guest in *The Mill
on the Floss.* As a lover and husband for Dorothea, he
appears quite inadequate; the difficulty is in seeing what
Dorothea and her author could see in him. One is left
feeling that he would be more convincing for the addition of a few spots of commonness.

One is bound to approach *Middlemarch* from the four

main stories that compose it. But not to be separated from them is the network of relations so closely woven among them that it makes the novel unrivalled in English as a panorama of provincial society at a specific date in history. As a panoramic novel its scope is vast; there is nothing, one feels, of the time of its action that is omitted: the struggle for Parliamentary reform, advances in medicine, the revolution caused by the coming of the railways, improvements in farming methods and rural housing—all are here. All the contrasting milieus are drawn with equal ease and authority, those of the country gentry and the bourgeois interiors of the Vincys and Bulstrodes alike. And the gallery of characters is as vast as the scene, far too big to be enumerated. One remembers especially, perhaps, Celia Brooke, her husband Sir James Chettam, the Cadwalladers, Mr. Brooke, the Vincys; but there is also the roll of Middlemarch doctors and lawyers and tradesmen. All, however small, start from the page in vivid and authentic life, as, for instance, the auctioneer Mr. Trumball, who for all the paucity of his appearances in the action astonishes whenever he does appear. He is a masterpiece of comic observation.

8

SERIAL PUBLICATION OF *Daniel Deronda* began on February 1, 1876. On May 27, Trollope wrote to Mary Holmes:

> Daniel D. has been a trying book to me. You perhaps know how much I love and admire her. She is to me a very dear friend indeed. It is so, whether I have told you so or not. But I think D. D. is all wrong in art. Not only is the oil flavoured on every page, (which is a great fault)—but with the smell of the oil comes so little of the brilliance which the oil should give! She is always striving for effects which she does not produce. All you say of Gwendoline's character is true. She disgusts, and does not interest,—as a woman may even though she disgusts. But Homer was allowed to nod once or twice, & why not the author of Adam Bede and Romola?

It was on the character of Gwendolen Harleth that public attention mainly seized, as George Eliot ruefully admitted when, in the following October, after the novel had been published in book form, she complained in a letter of "readers who cut the book up into scraps, and talk of nothing in it but Gwendolen," adding, "I meant everything in the book to be related to everything else there."

But by then, Henry James had published his *Daniel Deronda: A Conversation* in the New York *Nation*. It remains a fine piece of criticism and is interesting besides because the form allows James to express opposed contemporary views of George Eliot's work and of *Daniel Deronda* in particular and thereby gives him the opportunity of mediating between them.

There are three characters in the conversation: Theodora, who is serious and intense; Pulcheria, who as her name suggests is frivolous; and Constantius, a young book-reviewer who, like James himself at this time, has written a single novel. Pulcheria is satirical; for her, Gwendolen is "a second-rate English girl who got into a flutter about a lord"; she is "vulgarly, pettily, drily selfish," while Deronda is "a dreadful prig," without blood in his body. Theodora, on the other hand, is ecstatic in her enthusiasm; she finds Klesmer "almost Shakespearean" and Deronda "the most irresistible man in the literature of fiction." Through this clash of opposites, the judicious Constantius steers a middle course. He thinks they are "both in a measure right." Gwendolen is a "masterpiece," Grandcourt "a consummate picture of English brutality refined and distilled (for Grandcourt is before all things brutal)." Compared with these, he finds the Jewish part of the novel wearisome, and he distinguishes between the characters based on observation and the characters based on invention, characters he links with "two very distinctive elements" he discovers in George Eliot—"a spontaneous one and an artificial one."

Constantius's view of *Daniel Deronda* is very much the modern view. It has been given magisterial expression in F. R. Leavis's brilliant chapters on George Eliot in *The Great Tradition*, chapters which first appeared in 1945 and 1946 in *Scrutiny*. Leavis here more or less extracts another novel from *Daniel Deronda*—the "good part" of it, which he calls *Gwendolen Harleth*. And, incidentally, it is to Leavis, who printed it as an appendix to *The Great Tradition*, that most of us are indebted for our first acquaintance with James's imaginary conversation on *Daniel Deronda*.

Our knowledge of George Eliot's life shows how right James was in the distinction he made between the characters in the novel based on observation and those based on invention. The novel, as the experience of all who have read confirms, does fall into two parts, Gwendolen being at the centre of one and Deronda at the centre of the other. Gwendolen came from observation that fired the author's imagination. We can even put a date to the first spark. In October, 1872, when George Eliot and Lewes were staying at Homburg, in Germany, she wrote to Mrs. Cross, the mother of her future husband:

> The air, the waters, the plantations here, are all perfect— "only man is vile." I am not fond of denouncing my fellow-sinners, but gambling being a vice I have no mind to, stirs my disgust even more than my pity. The sight of the dull faces bending round the gaming-tables, the raking up of the money, and the flinging of the coins towards the winners by the hard-faced croupiers, the hateful, hideous women staring at the board like stupid monomaniacs—all this seems to me the most abject presentation of mortals grasping after something called a good that can be seen on the face of this little earth. Burglary is heroic compared with it. I get some satisfaction in looking on from the sense that the thing is going to be put down. Hell is the only right name for such places.

There, surely, we have the genesis of the superb opening chapter in *Daniel Deronda*. But, more or less by the same post, she was writing to Blackwood, her publisher:

> There is very little dramatic *Stoff* to be picked up by watching or listening. The saddest thing to be witnessed is the play of a young lady, who is only twenty-six years old, and is completely in the grasp of this mean, money-making

demon. It makes me cry to see her young fresh face among the hags and brutally stupid men around her.

There, surely, is the genesis of Gwendolen Harleth herself.

The Jewish side of the novel is another matter; this proceeds from invention, not from observation, from the artificial element in George Eliot rather than the spontaneous. She had not always admired or liked Jews. One finds her as a young woman in Coventry reacting strongly against Disraeli's *Tancred:*

The fellowship of race, to which D'Israeli so exultingly refers the munificence of Sidonia, is so evidently an inferior impulse, which must ultimately be superseded, that I wonder that even he, Jew as he is, dares to boast of it. My Gentile nature kicks most resolutely against any assumption of superiority in the Jews, and is almost ready to echo Voltaire's vituperation. I bow to the supremacy of Hebrew poetry, but much of their early mythology, and almost all their history, is utterly revolting. Their stock has produced a Moses and a Jesus; but Moses was impregnated with Egyptian philosophy, and Jesus is venerated and adored by us only for that wherein He transcended or resisted Judaism. The very exaltation of their idea of a national deity into a spiritual monotheism seems to have been borrowed from the other oriental tribes. Everything specifically Jewish is of a low grade.

It is, of course, quite likely that at that time and at that place she had not even met a Jew, but one guesses that the change in her attitude towards Jews was connected with her passionate enthusiasm for Spinoza, which Lewes shared. Lewes, indeed, had been led to Spinoza by a Jew—significantly, in the light of *Daniel Deronda,* named Cohn and, equally significantly, a watchmaker, like Mordecai. By the end of the eighteensixties, too, she and Lewes were close friends of Emmanuel Deutsch, an Orientalist who in 1869 became the leader of a Jewish "Back to Palestine" movement, in other words a pioneer of the Zionist movement.

As propaganda for Zionism, *Daniel Deronda* seems to me still eloquent and even moving, as Disraeli's special pleading for the Jews no longer is, and of real historical interest. When published, it had, of course, great topical appeal, and I don't think George Eliot should be robbed of the credit for her prevision. However remote it may have seemed in 1876—and obviously James in his imaginary conversation did not take the prospect of it very seriously—today Israel exists; and it might be noted, too, that Deronda, at the working-men's club at The Hand and Banner, prophesies "a great outburst of force in the Arabs, who are being inspired with a new zeal." This much can certainly be said: in *Daniel Deronda*, George Eliot showed a remarkable grasp of the meaning and the possibilities of nationalism. Whatever she was not doing when she was writing the Jewish parts of *Daniel Deronda*, she was certainly thinking.

All the same, the difficulty remains. On the one hand, there is that part of the novel which is the product of observation transfigured by imagination; on the other, the part that proceeds from invention, which is *voulu*. And what is still not easy to come to terms with is the discrepancy in quality between the two parts, which is so enormous as to make a positive gulf. For the two parts— one of which, for the sake of convenience, one might call, following Dr. Leavis, *Gwendolen Harleth*—are utterly dissimilar in kind. *Gwendolen Harleth* is a modern novel. It was modern when it was written, the only quite modern novel George Eliot wrote. The events narrated take place in 1860; which means that, at a blow, George Eliot cut herself off from what had been before a large part of her strength—the rendering of a period that was past and had, as it were, built-in resources of humour and nostalgia simply because it was past. Nor are the scene and society depicted provincial. They are, indeed, cosmopolitan and largely upper class, with no more place in them for the Vincys and Dodsons than for the Dolly Winthrops.

But *Gwendolen Harleth* is modern in another and more important sense. It is modern as *Madame Bovary* and *The Portrait of a Lady* are modern. Action derives solely from character; there is an almost entire absence of plot in the old-fashioned sense. Character is destiny, and implicit in Gwendolen's is both her hubris and her nemisis. Nothing could be cleaner, sparer, more austere than the narrative line of *Gwendolen Harleth*.

By contrast, what remains of the novel, the other half, which we may perhaps call *Deronda*, is old-fashioned in the extreme, a farrago of mysteries and melodrama, with Deronda uncertain of his family and birth, rescuing Mirah from suicide by drowning as she despairs of finding her long-lost mother and sister, discovering by pure chance her brother in Mordecai and, in the end, discovering his own mother and his race. As narration, it is so unconvincing that one finds it impossible to believe it was ever meant to be convincing.

This general air of unconvincingness invades even the rendering of the characters at the most literal level. There is, in *Gwendolen Harleth*, the musician of genius, Klesmer. He is German. An educated man, he speaks in nothing so crude as broken English; we are not in the least invited to laugh at his English, but his English, for all that, is established as a foreigner's. In *Deronda*, Mirah and Mordecai are Central European Jews, but no one would know this from the language George Eliot puts in their mouths. They speak perfect English. It is true that, before coming to England, Mirah has been an actress in the United States, which could account for her mastery of the language. No such explanation is available for Mordecai; and yet he commands in his adopted, and one would assume alien, tongue the eloquence of a Shelley.

It is as though probability is the last consideration George Eliot has had in mind. There is, too, another difference between the two parts of the novel, the difference in her attitude towards the characters of the two parts. In *Gwendolen Harleth,* there is not a personage who escapes her astringent irony of presentation. We feel, in consequence, that she has seen all round them, knows them inside out, as Jane Austen normally knows hers. Indeed, her irony here is closely akin to Jane Austen's. It is not destructive; as Leavis remarks, of George Eliot's presentation of the Reverend Mr. Gascoigne, Gwendolen's uncle, it is not satire; there is no intent to expose hypocrisy, for hypocrisy is not there:

> This match with Grandcourt presented itself to him as a sort of public affair; perhaps there were ways in which it might even strengthen the Establishment. To the Rector, whose father (nobody would have suspected it, and nobody was told) had risen to be a provincial corn-dealer, aristocratic heirship resembled regal heirship in exempting its possessor form the ordinary standard of moral judgments. Grandcourt, the almost certain baronet, the probable peer, was to be ranked with public personages, and was a match to be accepted on broad general grounds national and ecclesiastical.

Irony such as that is the index of the author's knowledge of his character and the character's environment. It gives us, to use George Eliot's own words, both the medium in which the character moves and the character itself. It presents the moral assumptions which govern the character's behaviour. It does not in the least preclude recognition that, as men go, the character, in this instance the Reverend Mr. Gascoigne, is a good man. What it does, however, is to put the kind of goodness that he has, the moral assumptions that govern him, into perspective.

Irony of this kind is essentially critical, and none of the characters in *Gwendolen Harleth* escape its unfaltering illumination.

George Eliot's critical irony is, as it were, switched off at the main as soon as she moves from *Gwendolen Harleth* to *Deronda*. In the latter, we are in a world which—quite literally, from its author's point of view—moves to entirely different laws, whose inhabitants are rendered not with the impartiality of full knowledge—which, because it is full, can make allowances and accept and forgive—but, instead, as figures to be warmly loved or utterly detested at, so to speak, the author's word.

Take, for example, the Meyricks, that family of busy lady-artists happy in their poverty off the Brompton Road, to which Deronda takes Mirah for shelter. James's Theodora, in his dialogue, finds them "delicious," Pulcheria agrees that they are the best thing in the book, and for Constantius they are "delightful"; indeed, the merit of the Meyricks is the one thing they are unanimous on. They seem now to be conceived and executed in excessive sweetness, wholly sentimental notions of human beings and of family life.

Much better, in fact the best part of *Deronda*, is the delineation of the Cohens, the Jewish pawnbroker's family. Of Mordecai and Mirah, it seems to me nothing can be said at all; as human beings they are simply not there. Deronda himself is rather different. He is an intellectual conception, but there are moments when he is buttressed, as it were, as Tito in Romola is, by his author's analysis of him; and in the beginning, he is surrounded with the aura of mystery he has for Gwendolen. As soon as he is presented full face, however, mystery in any but a mechanical sense departs, and he becomes, for all the care lavished on him, a puppet in a complicated plot that is the more incredible the more it unwinds.

So in *Daniel Deronda* we seem to have two disparate

novels—one thoroughly modern, inducing George Eliot
to a kind of achievement she had never attained before,
the other Victorian in the bad sense. They meet, of
course, in Deronda, which is why any actual attempt to
extrapolate a novel *Gwendolen Harleth* from the work
as a whole would be impossible. As we have seen,
George Eliot herself "meant everything in the book to be
related to everything else there," and so in a way it
is. But the way or the ways are pretty faint.

There are, for instance, parallel situations; the circumstances of the Davilows in essence correspond more
or less to those of the Meyricks. There are attempts, too,
at a unifying imagery. Gambling becomes a powerful
symbol of a wrong way of life, and the gambling motif
with which the novel opens is repeated at the end in
the figure of Mirah's and Mordecai's father. In *The Art of
George Eliot,* W. J. Harvey works out in considerable
detail the manifold ways in which George Eliot uses
imagery derived from art, and especially from the theatre,
as she pursues the theme of illusion and reality, particularly in relation to Gwendolen. There are parallels, cor-
respondences, here, too: Mirah has been an actress;
Gwendolen thinks she might very easily become one,
until she is abruptly wakened from her dreams of effort-
less and rapid success by Klesmer.

Looked at from this point of view, Klesmer, who, as
many critics have said, is one of the few convincing
representations of genius in fiction, exists in correspon-
dence to Deronda; in relation to Gwendolen, they may
be said to stand for alternative absolutes: the demands of
art and the demands of the life of action. The reality
of the life of art as exemplified in Klesmer, both as a
character and in the speeches he makes in his confrontation with Gwendolen when she seeks his advice on how
to become an actress, is infinitely more convincing than
that of the life of action as exemplified in Deronda. And

to say this is merely to say again that George Eliot's
strength in *Daniel Deronda* is in direct proportion to the
degree in which she is writing from observation.

This means that *Gwendolen Harleth*—and *Gwendolen
Harleth* is superb—is one of the high peaks of fiction.
One cannot say it is better than *Middlemarch;* it is different—in essence, a wholly new venture, which could not,
I think, be predicted from anything George Eliot had
written before. Except as anti-type, Gwendolen herself
is not the young woman one associates with George Eliot,
and it is, one feels, a sign of prodigious development on
George Eliot's part that she can create her. She is cold,
calculating, arrogant, self-willed, where what one thinks
of as the typical George Eliot young women, Maggie and
Dorothea, are warm, impulsive, self-sacrificing to the
point of masochism. What she has in common with them
is intelligence, but she has, as well, a flashing brilliance,
a wit, an inordinately high degree of self-valuation as a
unique being, reminiscent of Jane Austen's Emma. These
are factors that remove her from anything but remote
kinship with Rosamund Vincy. There are faint adumbrations of her in Esther Lyon in her earlier manifestations;
but, if we are to seek analogues for her in George Eliot's
work, she is much nearer Mrs. Transome, for she is a
tragic character. It is the index of George Eliot's success
with her that we accept her as such.

She is, as the title of the first book indicates, a spoilt
child. She has, when we first meet her, an "implicit
confidence that her destiny must be one of luxurious
ease," a confidence that comes in part from "that sense of
super claims which made a large part of her consciousness." She has—and here again George Eliot reminds us
of Jane Austen—been badly brought up; she had lacked
any proper instruction in religion. Throughout her life
she has been given way to:

Having always been the pet and pride of the household, waited on by mother, sisters, governess, and maids, as if she had been a princess in exile, she naturally found it difficult to think her own pleasure less important than others made it, and when it was positively thwarted felt an astonished resentment apt, in her cruder days, to vent itself in one of those passionate acts which look like a contradiction of habitual tendencies. Though never even as a child thoughtlessly cruel, nay, delighting to rescue drowning insects and watch their recovery, there was a disagreeable remembrance of her having strangled her sister's canary-bird in a final fit of exasperation at its shrill singing which had again and again jarringly interrupted her own. She had taken pains to buy a white mouse for her sister in retribution, and though inwardly excusing herself on the ground of a peculiar sensitiveness which was a mark of her general superiority, the thought of that infelonious murder had always made her wince. Gwendolen's nature was not remorseless, but she liked to make her penances easy, and now that she was twenty and more, some of her native force had turned into a self-control by which she guarded herself from penitenial humiliation. There was more show of fire and will in her than ever, but there was more calculation underneath it.

But "spoilt child" by no means alone sums her up. Her brilliance and self-possession, which are manifest in every encounter she has with those about her, are such as to make it quite possible to see her as "a princess in exile." The truth is, as George Eliot demonstrates by the words she puts into her mouth and the behaviour she ascribes to her, Gwendolen dominates the social scenes in which she habitually moves precisely because she is a superior being. Indeed, if this were not established we should not be willing to accept her in the end as a tragic figure, as we unhesitatingly do.

"Always she was the princess in exile, who in time of famine was to have her breakfast-roll made of the finest-

bolted flour from the seven thin ears of wheat, and in a
general decampment was to have her silver fork kept
out of the baggage." That is how she sees herself. George
Eliot's irony provides the comment and indicates her
besetting weakness, which, in fairness to her, it must be
said is one—in part, at any rate—inseparable from her
position as an upper-class young Englishwoman of her
period. Her besetting weakness is not selfishness or
calculation but ignorance, ignorance of the world outside
her small family and social circle. "Gwendolen's confidence lay chiefly in herself. She felt well equipped for
the mastery of life." This is an illusion: she simply does
not know enough; it is impossible that she should—and,
besides, she lacks humility. "Mamma, I see now why girls
are glad to be married—to escape being expected to please
everybody but themselves.'" Her ignorance, together
with her self-regard—and the two are hardly to be
distinguished, so much is she the princess of the tiny
world that is her only experience of life—leads her to
take it for granted that, within a matter of months, she
will be acclaimed as a great actress or a great singer;
nothing, until Klesmer disabuses her, is beyond her
powers.

The clue to her is to be found, perhaps, in the thought
that comes into her mind when she first sees Grandcourt
at the archery meeting at Brackenshaw Park: "He is not
ridiculous." The inference is plain: Grandcourt is unique
among all the people she had previously met, Klesmer
excepted, in not being ridiculous. And Grandcourt is as
remarkable a portrayal as Gwendolen. James's Constantius, having pointed out that he is "before all things
brutal," describes him as a "consummate representation of
the most detestable kind of Englishman—the Englishman
who thinks it low to articulate." He is right; and yet
there is more to Grandcourt than this. He is, of course,
a man as seen by a woman novelist; a male novelist,

James himself, would have seen and portrayed him
differently and, in doing so, would have robbed him of
a certain sexual magic. As it is, like Charlotte Brontë's
Rochester, he has the demonic quality of the incompletely
understood. We, as readers of the omniscient novelist,
are told everything we need to know about him; he is
conveyed to us in ironical, economic detail, in terms of his
attitude towards his dogs, his insolent superiority to
Lush, his massive omnipresent boredom. The remarkable
thing is that he is not a caricature. He is, in fact, anything but a caricature; he has, and impresses on us, the
complete assurance of the self-contained man; and though
he is before all things brutal, he also has tremendous
style.

It is this that is his attraction for Gwendolen, who
necessarily knows so much less about him than we do.
For her, he is the complete formal man, whose style, in
the sense of habitual manner of behaviour, is so much of
a piece as to impose itself without criticism on all about
him. It is the behaviour of a man to whom nobody else
and nothing else exist except as attributes of himself,
as things to be used. It is the behaviour, in other words,
of the masculine counterpart of the woman that Gwendolen aspires to be. But Grandcourt is her nemesis. All
the advantages in their marriage are his—wealth, position,
age, experience of the world, the very fact of being a
man and the kind of man he is: "If this white-haired man
with the perpendicular profile had been sent to govern a
difficult colony, he might have won reputation among
his contemporaries." It is Gwendolen who becomes the
possession, the thing, the attribute:

> This beautiful, healthy young creature, with her two-and-
> twenty years and her gratified ambition, no longer felt
> inclined to kiss her fortunate image in the mirror; she looked
> at it with wonder that she could be so miserable. One belief
> which had accompanied her through her unmarried life as

a self-cajoling superstition, encouraged by the subordina-
tion of every one about her—the belief in her own power of
dominating—was utterly gone. Already, in seven short
weeks, which seemed half her life, her husband had gained
a mastery which she could no more resist than she could
have resisted the benumbing effect from the touch of a
torpedo. Gwendolen's will had seemed imperious in its
small girlish sway; but it was the will of a creature with a
large discourse of imaginative fears: a shadow would have
been enough to relax its hold. And now she found a will
like that of a crab or a boa-constrictor which goes on
pinching or crushing without alarm at thunder. Not that
Grandcourt was without calculation of the intangible effects
which were the chief means of mastery; indeed he had a
surprising acuteness in detecting that situation of feeling
in Gwendolen which made her proud and rebellious spirit
dumb and helpless before him.

As she reflects:

"He delights in making the dogs and horses quail: that
is half his pleasure in calling them his. . . . It will come to
be so with me; and I shall quail . . ."

From beginning to end, the scenes between them are
masterly and done with wonderful economy; witness the
proposal scene, in which Gwendolen accepts Grandcourt
against all her avowed intentions:

"You accept my devotion?" said Grandcourt, holding his
hat by his side and looking straight into her eyes, without
other movement. Their eyes meeting in that way seemed to
allow any length of pause; but wait as long as she would,
how could she contradict herself? What had she detained
him for? He had shut out any explanation.
"Yes," came gravely from Gwendolen's lips as if she had
been answering to her name in a court of justice. He received it gravely, and they still looked at each other in the
same attitude. Was there ever before such a way of accepting the bliss-giving "Yes"? Grandcourt liked better to

be at the distance from her, and to feel under a ceremony
imposed by an indefinable prohibition that breathed from
Gwendolen's bearing.

But he did at length lay down his hat and advance to
take her hand, just pressing his lips upon it and letting it
go again. She thought his behaviour perfect, and gained
a sense of freedom which made her almost ready to be
mischievous. Her "Yes" entailed so little at this moment,
that there was nothing to screen the reversal of her gloomy
prospects: her vision was filled by her own release from the
Momperts, and her mother's release from Sawyer's Cottage.
With a happy curl of the lips, she said—

"Will you not see mamma? I will fetch her."

"Let us wait a little," said Grandcourt, in his favourite
attitude, having his left forefinger and thumb in his waist-
coat-pocket, and with his right caressing his whisker, while
he stood near Gwendolen and looked at her—not unlike a
gentleman who has a felicitous introduction at an evening
party.

"Have you anything else to say to me?" said Gwendolen
playfully.

"Yes.—I know having things said to you is a great bore,"
said Grandcourt, rather sympathetically.

"Not when they are things I like to hear."

"Will it bother you to be asked how soon we can be
married?"

"I think it will, today," said Gwendolen, putting up her
chin saucily.

"Not today, then, but tomorrow. Think of it before I
come tomorrow. In a fortnight—or three weeks—as soon as
possible."

"Ah, you think you will be tired of my company," said
Gwendolen. "I notice when people are married the husband is not so much with his wife as when they were
engaged. But perhaps I shall like that better too."

She laughed charmingly.

"You shall have whatever you like," said Grandcourt.

"And nothing that I don't like?—please say that; because
I think I dislike what I don't like more than I like what I

like," said Gwendolen, finding herself in the woman's paradise where all her nonsense is adorable.

Grandcourt paused: these were subtleties in which he
had much experience of his own. "I don't know—this is
such a brute of a world, things are always turning up that
one doesn't like. I can't always hinder your being bored.
If you like to hunt Criterion, I can't hinder his coming
down by some chance or other."

"Ah, my friend Criterion, how is he?"

"He is outside: I made the groom ride him, that you
might see him. He had the side-saddle on for an hour or
two yesterday. Come to the window and look at him."

They could see the two horses being taken slowly round
the sweep, and the beautiful creatures, in their fine groom-
ing, sent a thrill of exultation through Gwendolen. They
were the symbols of command and luxury, in delightful
contrast with the ugliness of poverty and humiliation at
which she had lately been looking close.

That is great writing. And with what consummate
skill are we prepared for Gwendolen's acceptance of
Grandcourt, against her deepest feelings, her best intentions, her promise to Mrs. Glasher. She is pinned
down by intolerable pressures—by her mother's sudden
fall into poverty, by her bitter humiliation at the hands
of Glesmer, by the hated alternative to marriage with
Grandcourt—to go as governess to Bishop Mompert's
children—by, indeed, the very desirability of the match
in the eyes of the world, as represented by her uncle the
rector. After all, she alone of her circle knows of the
existence of Mrs. Glasher and Grandcourt's illegitimate
children. Her tragedy is that it is only after her self-
betrayal, marriage to Grandcourt, that she discovers, in
continual contact with his "refined negations," that she
has "a root of conscience in her" that makes her pray,
"I will not mind if I can keep from getting wicked." Her
punishment is in her symbolic murder of Grandcourt, as
she confesses it to Deronda:

"I saw him sink, and my heart gave a leap as if it were going out of me. I think I did not move. I kept my hands tight. It was long enough for me to be glad, and yet to think it was no use—he would come up again. And he *was* come—farther off—the boat had moved. It was all like lightning. 'The rope!' he called out in a voice—not his own—I hear it now—and I stopped for the rope—I felt I must—I felt sure he could swim, and he would come back whether or not, and I dreaded him. That was in my mind— he would come back. But he was gone down again, and I had the rope in my hand—no, there he was again—his face above the water—and he cried again—and I held my hand, and my heart said, 'Die!'—and he sank; and I felt 'It is done —I am wicked, I am lost!'—and I had the rope in my hand— I don't know what I thought—I was leaping away from myself—I would have saved him then. I was leaping from my crime, and there it was—close to me as I fell—there was the dead face—dead, dead. It can never be altered. That was what happened. That was what I did. You know it all. It can never be altered."

Yet George Eliot's triumph with Gwendolen and Grandcourt is hardly to be separated from her general triumph with the scenes in which they appear, alone or together—the scene in the casino with which the novel opens, the archery meeting, the scenes between Grandcourt and Lush, a brilliant characterisation; and her hand is just as sure in episodes in which neither is present, as, for instance, in Klesmer's proposal to Catherine Arrowpoint and the Arrowpoints' reaction to it. All have a directness, a magisterial authority, surpassing anything she had done before, even in *Middlemarch*. She is in complete control of the world she is describing. She knows it through and through, as well, one feels, as Jane Austen does hers in *Emma* or Flaubert his in *Madame Bovary*, and she renders it as perfectly as they do theirs.

G EORGE ELIOT DIED AT 61, not a great age, and, as
Daniel Deronda shows, her art was widening,
deepening, growing in authority until the end; so much
so that one feels that had she lived another ten years,
she might have surpassed even the finest of the novels
she left behind. This sense of incomplete fulfilment of
her genius comes in part from the fact that she was
writing during a period of transition in the novel, so that
constantly one finds oneself judging her work by two
different and distinct standards: that by which one
judges her English contemporaries, such as Thackeray
and Trollope, and that invoked by her European contemporaries, Flaubert and Turgenev in particular.

In the purely English context of Victorian fiction,
George Eliot is obviously a great writer, second only to
Dickens, who admittedly is very great by criteria that
transcend the temporal and local. Seen thus, the books
we shall almost certainly value the most are the early
ones, *Adam Bede, The Mill on the Floss* and *Silas
Marner,* as Trollope did; and we shall probably reserve
our especial admiration for the rustic characters like Mrs.
Poyser and the entire population of Raveloe, for the
small-town characters like the Dodson sisters and their
husbands and, of course, Maggie Tulliver and her family.
These are indeed splendid achievements. As a delineator

of English provincial, and particularly rural, life at one
stage in its history, the years immediately before the
coming of the railways, she has no equal except Thomas
Hardy.

If we set her, however, side by side with her European
contemporaries, the emphasis falls rather differently,
and it is the last novels, *Felix Holt*—if only for the tragic
figure of Mrs. Transome—*Middlemarch* and the Gwendolen Harleth parts of *Daniel Deronda,* that become of
most value. These novels are distinguished by a psychological realism unknown before in English fiction except,
perhaps, in Fielding's *Amelia* and Jane Austen's *Emma.*
Character has become destiny, and character is subjected
to intense critical scrutiny. At the same time, the effects of
environment on character are studied with such minute-
ness that she can almost be seen as a fore-runner of
naturalism, abhorrent though naturalism would have
been to her as an aesthetic doctrine.

And though she lacked the artistry of Flaubert and
Turgenev, partly because of the tradition of fiction in
which she grew up, partly because of her ethical pre-
occupations, she seems to me inferior to neither in
analysis of character or in the rendering of specific social
scenes. Her triumph in these respects is in a way the
more impressive because, in contrast with Flaubert and
Turgenev, her prose is not distinguished and seldom is
more than adequate. It lacks the graces that we are
often pleased to call superficial but that we invariably
respond to with delight whenever we find them, especially
when they are combined with the high seriousness of
novelists like Flaubert and Turgenev.

It is, as E. M. Forster notes in *Aspects of the Novel,* on
her massiveness that she depends—"she has no nicety of
style." Intelligent readers—among them George Gissing,
who was not a frivolous man or novelist—have found
her massiveness "heavy"; and Gissing, indeed, found her

style in *Daniel Deronda* "monstrous." He was wrong, but one sees what he means. There is a formidableness in George Eliot, almost an over-anxiety to put over her moral seriousness, that can be daunting. It was part of her temperament; it was accentuated by her early experiences and by the findings of her intellect; it was inseparable from her. The style was, in fact, the woman. In the end, it must be taken—or left—as every writer's style must be.

Her influence is difficult to assess because she has been for almost a century part of the climate in which every English novelist has been brought up, one of the permanent and principal standards of reference for the writing of novels. Her influence on some authors is plain; on others, less certain.

Mr. Leavis has said that Thomas Hardy got his prompting to the Wessex novels from *Silas Marner* and his use of the old Saxon place-name of Wessex from *Daniel Deronda*. None of this, it seems to me, is proved. Hardy first used the word "Wessex" in *Far from the Madding Crowd*, which appeared four years before *Daniel Deronda;* and, in any case, the old place-name had been given new currency some years before by the Dorset poet William Barnes, the first poet Hardy knew in the flesh. George Eliot herself Hardy admired as a "great thinker," one of the greatest of living writers, "though not a born story-teller by any means." Their notions of the art of story-telling were, in fact, poles apart.

They had certain things in common which make them appear closer to each other than they actually were. Though born in very different parts of England, with twenty years between them, both spent their childhood in the pre-railway countryside and looked back to it with nostalgia when they began to write. Both, though intellectually advanced thinkers—they had come under similar influences—were in some ways naturally Tory and naturally Christian. They were sufficiently similar for reviewers to attribute *Far from the Madding Crowd* to George Eliot when it appeared anonymously in serial form in *Cornhill* in 1874. Yet the differences between them are as real as the affinities, which are largely those of circumstance and time. Hardy was no determinist as George Eliot was; indeed, his world and the fate of his characters are governed by chance, which George Eliot explicitly rules out as an operative factor in the determining of her characters' lives. Hardy had no need of George Eliot to teach him a tragic view of life—he got it from the Greeks. And the glaring difference between them may be seen at a glance if one contrasts Hardy's Tess with Eliot's Hetty.

They seem at their closest in their rendering of rural characters, that part of Hardy for which Leavis thinks he was indebted to George Eliot; and we inevitably set the Hardy of *Under the Greenwood Tree* and of the rustic choruses throughout most of the novels side by side with the rustics of *Adam Bede* and *Silas Marner*. But Hardy himself found George Eliot unconvincing as a creator of country folk; hers talked, it seemed to him, like small-town people. The truth is, where their rustic characters are concerned, both belonged to the same tradition of English writing: behind both are the country clowns of Shakespeare and Fielding and the peasants of Scott.

Elsewhere, George Eliot's influence is unequivocal. She was a constant preoccupation of Henry James, who wrote ten reviews or essays about her, and her presence is to be felt behind many of his novels. Leavis has shown, it seems quite clear to me, the debt *The Portrait of a Lady* owes to the Gwendolen Harleth part of *Daniel Deronda;* and beyond this, W. J. Harvey is surely right when he suggests that in the novels from *The Portrait of a Lady* to *The Tragic Muse*, James is more like George Eliot than he is like any modern novelist.

Her influence on D. H. Lawrence was only slightly less obvious, though of a different order. It was, one guesses, twofold. One part stemmed from the situation Lawrence found himself in as a provincial youth living far from the conventional literary scene. George Eliot had come from a provincial area not very far from his own and had written, in *Adam Bede,* of a countryside he knew. This in itself was enough to make him adopt her as a literary ancestor, as someone to be emulated because of the very similarity of their situations. But also, for Lawrence, George Eliot was the founder of the modern novel; as he wrote in an early letter, she was the first novelist to start "putting all the action inside," and when he began to write his first novel, *The White Peacock,* it seems to have been with George Eliot, and with *The Mill on the Floss* particularly, in mind. But even apart from this, when all allowance has been made for the great differences in the nature of their genius, her presence can be sensed behind the pages of the first part of *The Rainbow,* in which, as a delineator of rural life, he is not her inferior. Lawrence was steeped in her, and it is this that one feels in his work rather than specific influences, though Graham Hough has observed, in his book *The Dark Sun,* a specific indebtedness to the Dorothea-Casaubon relationship in Lawrence's short story *Daughters of the Vicar.*

But another novelist, very different from Lawrence, was also steeped in George Eliot. "Two pages of *The Mill on the Floss* are enough to start me crying," Marcel Proust wrote; and from boyhood, it had been one of his favourite books. André Maurois has surmised that it must have been much in Proust's mind when he was planning *À la recherche du temps perdu* and points out the startling resemblance between its opening paragraphs and those of *Du côté de chez Swann.*

In England, at present, there is perhaps a tendency

slightly to over-estimate George Eliot, possibly because she is the one English novelist of her time, Dickens apart, who can be set against her great European contemporaries in seriousness and also because of her special position in the history of English fiction.

Thus, the contemplation of George Eliot leads two such utterly dissimilar critics as David Cecil and F. R. Leavis to invoke Tolstoy. Cecil calls *Middlemarch* "a provincial *War and Peace*," while Leavis, thinking primarily of *Daniel Deronda*, writes, "Of George Eliot it can . . . be said that her best work has a Tolstoyan depth and reality." When Leavis says that "the extraordinary reality of *Anna Karenina* . . . comes of an immense moral interest in human nature that provides the light and courage for a profound psychological analysis" and applies this to *Daniel Deronda*, he is obviously right, just as, if you are looking for an English counterpart to *War and Peace*, *Middlemarch* is probably the nearest thing you will find.

Yet, though the comparison with Tolstoy and the praise it implies are just, it needs great qualification if it is to make complete sense. One may compare George Eliot to Tolstoy—one would never dream of comparing Tolstoy to George Eliot; for there are whole vast ranges of human experience that Tolstoy has which are not to be found in George Eliot at all. As W. J. Harvey says, commenting on Leavis's comparison of *Daniel Deronda* with *Anna Karenina:* "One cannot imagine George Eliot encompassing either Levin's simple joy at being alive and in love or the complex intensities of Anna Karenina's passion." In the same way, *Middlemarch* is not only much smaller, much more restricted, than *War and Peace* as a panorama of life in history, it also lacks entirely the simple, sensuous, almost animal joy in being alive that permeates Tolstoy's novel.

There is something else, too, beside the fact (which

English critics tend to forget) that when we contemplate George Eliot's novels, we are contemplating works that, *Silas Marner* and *Middlemarch* apart, are much more deeply flawed than we expect great novels normally to be: the best parts of *Felix Holt, the Radical* and *Daniel Deronda* may touch the heights of fiction, but they are only parts, for all that, and they cannot be extrapolated from the books that contain them and be made into independent novels at the critic's whim. V. S. Pritchett has said, "There is no real madness in George Eliot," meaning by madness the sense of the forces of the irrational, the daimonic, the incomprehensible in men's lives. This is to say that, for all her profound reverence, she is never a religious novelist; and the reality of sexual passion is also foreign to her. Her view of life was unswervingly and all the time a moral view, and nothing that existed outside the moral view, that could not be netted by it, existed for her. The loftiness of the moral view she held cannot prevent us thinking that there is a great deal in life that cannot be adequately explained or illuminated by it. The world is not primarily, as it too often seems in her novels, a gymnasium for the exercise and development of the moral faculties. Her picture of life, then, is more limited than her most fervent admirers are always willing to admit.

But her power and authority as a novelist come directly from the very passion with which she held to her moral view of life. Ben Jonson said of John Donne that he was "the first poet in the world in some things," and perhaps this is the highest praise that can be given to any imaginative writer except the very greatest, the transcendent half dozen or so. George Eliot is the first novelist in the world in some things, and they are the things that come within the scope of her moral interpretation of life. Circumscribed though it was, it was certainly not narrow; nor did she ever forget the difficulty attendant upon the moral life and the complexity that goes to its making. Indeed, it is the complexity that persuades us of its truth, the truth that reminds us that men are members one of another, linked to their fellows and to the past by innumerable fine filaments of work, habit and behaviour, to say nothing of kinship and duty.

"It is the habit of my imagination," she wrote, "to strive after as full a vision of the medium in which a character moves as of the character itself." Her novels satisfy both as detailed re-creations of specific environments and as collections of portraits of unique human beings. Because her characters are inescapably social beings—shaped by their own actions, admittedly, but shaped, too, by the pressures of the society in which they live and its history —for George Eliot, moral choice is never something in the abstract. It exists—and this is the hall-mark of her genius as a novelist—in the realm of the agonisingly personal. It is here, in her recognition of the individual's responsibility to his deepest sense of what is right, that her triumph as a novelist lies.